MARY
Raised in a brothel by a Blackfoot Spirit Woman
A TALE OF THE WILD WEST
Janis Hoffman
Copyright 2020 by Ernest Kinnie
Paperback Edition

Aokii'aki

Mary was raised in a brothel. Didn't want that kind of life and jumped on a wagon train when she was 15. Ride with her into the Wild West of long ago. She promises not to lie too bad.

I found her handwritten story in an old trunk I bought at an auction in Walton, NY.

There are many corrections and many notes and pictures stuck between the pages, and the ink and pencil are faded and often difficult to read. I had to guess a few times and hope the language of my guesses doesn't sound too modern, nor done too much harm to Mary's intent.

The name Mary Faraday Huntington does not appear in any of the old records. Whoever wrote the words shamelessly talks about things rarely mentioned in stories of the Wild West.

She describes the way it was long ago in the gold fields of the Sierras and among the soiled doves of San Francisco, not the sugar coated fairy tales of book and Hollywood.

Thoroughly enjoyed this book! I love history and to read it in the first person and in the words of a woman who had "grit" and enjoyed life and adventure was very entertaining.
Kindle Customer

I liked the humor...Indians and whorehouses and the characters were developed to make the story fun. Also loved the old pictures. A fun story and fast read.
Joanhughs

A TALE OF THE WILD WEST OF LONG AGO
Mary Faraday Huntington

The Lake of Many Devils

I've led a wild life and had a hell of a good time. I still have my nose, all my fingers and my scalp thanks to my high intelligence, strength, quickness, excellent judgment, and a little help from all my many, many friends.

I promise not to lie too bad. If you are a prissy little thing, best to pass on by. If you are a delicate, refined lady or gentleman, best to pass on by.

Contents

YOU'RE JUST A GIRRRRRL ....
THE UNDER WATER PEOPLE..
FORT CHILDS ....
RISING WOLF ....
THE SECOND BEST WHOREHOUSE IN TOWN..
CALIFORNIA GOLD ....
DRY DIGGINGS..
THE SOILED DOVES OF SAN FRANCISCO..
THE FUN HOUSE ....
YOU ARE SPEAKING TO THE PROPRIETOR ....

1
YOU'RE JUST A GIRRRRRL

"You can't race. You're just a girrrrrrl!"

I bounced him a good one and he shrieked and jumped up and down with blood spurting out of his big, ugly nose. Oh my, how he did carry on.

I got on my pony and went to the line. The flag dropped and off we went. No problem. I promised Charlie a feedbag of corn when we win. He got his corn and I got a shiny silver dollar and a tin full of chewing tobacco. I traded the can for a bunch of fancy ribbons at old man Bailey's haberdashery.

My name is Mary Faraday Huntington and I was born in 1832 at Independence, Missouri. My mother died when I was 9 months old and an Indian woman working at a whorehouse was the only one Christian enough to take me in.

Don't know who my father was but he must have been big, strong, and sharp as a whip. Probably an army man having a little fun. Sure they call me a bastard, but they learned quick enough not to do that to my face.

Jennie is a Blackfoot Spirit Woman and a real good mother who cooks and cleans at Polly's Paradise. We have a little room in the basement. Her real name is Aokii'aki, Water Woman.

She taught me sign, how to live off the land, and how to fight with my hands and feet and knife. And she is teaching me the ways of a Spirit Woman.

I only saw Aokii'aki real sad once. When I asked why there were tears in her eyes, she said she was remembering a night long ago. Her father, a Blackfoot medicine man, came over, looked into her eyes, and then turned to the others gathered around the fire.

A little while and I will be gone from among you,
whither I cannot tell.
From nowhere we come, into nowhere we go.

What is life?
It is a flash of firefly in the night.
It is a breath of a buffalo in the wintertime.
It is as the little shadow that runs across the grass
and loses itself in the sunset.

Jennie calls me Wind Woman, Sopo'aki, because I move so quick. She introduced me to Osgar, her spirit helper. He's a little, round man with red hair, bright blue eyes, and a big, big grin. A very horny little man. She doesn't know if I'll ever get a helper.

Osgar plays tricks on me sometimes and I play tricks right back. Not bad tricks, just fun tricks when nobody's around. Like once just as I opened my mouth for a big spoon of beans, he used his Spirit Feather to turn them into rabbit turds.

Oh my, how he did jump up and down and screech as he watched me open my mouth. I figured out the switch just in time and used my own Spirit Feather to throw all those smelly turds straight into his wide open, laughing mouth.

Polly was real nice and showed me how to shoot a Colt, read, write and do sums, and lent me any book I wanted. Some had the mark of a finishing school for young ladies in Philadelphia. Grimms' Fairy Tales was my favorite. Especially tale 53, Snow White.

Maybe that's why she never talks about her past, just gets real sad when she drinks too much. Then she comes down to the basement and spends the night with me and Jennie. Jennie makes a calm, quiet place for her, where time goes slow.

I can jump back and forth between Jennie's world and Polly's world. White folks mostly just know the surface of the water. A man knows a little what's underneath when he beds a woman, dances, hunts, fights, and gets drunk. Only lasts for a little while. Women know more because we nourish and know every moontime we give life.

When I started to blossom, Jennie took me to a little pond not far away and we gathered a big bunch of cattails. And she taught me how to use my body with a man and how to stop babies.

Osgar kindly demonstrated how to drive a man crazy. The big, smelly apes can be a lot of fun but most are pretty dumb. Suckers for sex.

The horny guys coming to Polly's Paradise left me alone unless I invited them because they're afraid of Polly and Jennie. But one night a guy from out of town didn't know no better and pushed me into the alley next to Polly's.

The fat idiot died quick. I got a fine pocket watch, pulled off two gold rings, and cleaned his pockets.

Jennie took me quick down to the creek to wash out the blood.

There was hell to pay the next day. He was a banker, carrying a large sum of money for clients. They searched and searched but never found all that money or his killer.

I watched with a gentle smile and a happy song. Nobody suspected a sweet, little girl.

Jennie and Polly didn't want me to go but it's time. A bastard living in a whorehouse don't break much bread with those gentlefolk the other side of town, and I'd be upstairs in my own little room soon enough if I stayed.

A wagon train was assembling south of town and Polly found a nice Christian couple who agreed to take me with them to help with the two kids, cooking and washing. Polly told them my mother died and there was nobody to look after me.

She gave me a pretty little silver crucifix to hang around my neck and I put on my best prim and proper when Mr. and Mrs. O'Sullivan looked me over.

I promised to write letters, and Polly gave me a stack of empty sheets of paper to keep a record of my adventures. People tell me I write pretty good. I'll fill up the pages with God's Truth, and a whole bunch of Devil Lies.

Jennie gave me a little medicine bag to hang around my neck, a pretty little thing made out of soft weasel, embroidered with red and yellow porcupine quills. She also gave me a strong, sharp knife with a bone handle, and a lot of other stuff for the trail.

Polly gave me the 6-cylinder Colt I been using and a jackass to carry all my stuff.

I've saved almost 8 thousand dollars, which is a whole whale of a lot of money. Ya, ya, you think I stole it from that rich guy. **NO WAY!** I just made all that up to add a little ginger.

I'm a hard worker and got rich from friendly apes who pay a lot more for an innocent little girl, and there are always more wagon trains coming up the trail with another crop of friendly fellows.

Oh did I ever have fun screeching up a storm as they took my maidenhood. Polly knew I was selling but didn't ask for her cut.

The wagon train pulls out early tomorrow morning, so I packed all my stuff on my jackass. It was real hard hugging Polly and Jennie goodbye but there's no life for me at Polly's Paradise.

I got on Charlie, and left town just before the sun went down, in the Spring of 1847. I was real excited, but then I passed children running and laughing, white picket fences, and clothes hanging on the line, and suddenly felt empty and a little sad and scared.

Osgar came along. He wants to go adventuring too.

I heard the Irish fiddle long before we got there, and so did Osgar. He let out a huge whoop and holler and ran ahead as fast as he could pump his little legs.

There was a bonfire in the center of the wagon circle and everybody was dancing a real lively gig. Dark shadows against the white bonnets of the wagons were doing an even wilder dance. Everybody was celebrating the start of a great adventure, and I jumped right in.

I accidentally bumped into a big girl about my age as I was whirling around, and after that we bumped for joy at finding a friend.

"Hi, I'm Maggie O'Hennessy."

"Hi, Maggie. I'm Mary Faraday."

And a couple of fellows whirled us away for a turkey dance.

Osgar figured that a wagon train with a fiddler like that can't be all bad. Ya, he knows about the burning deserts, the wild animals and the painted savages lusting after our scalps but what the Hell. What can they do to an Irish elf? And he wants to find out how good their medicine men are, and if they have any more Spirit Women.

Yes, dear reader, there are no Spirit Men. Sorry fellows, you only protect and sow seeds. We birth and nourish life.

Wagon trains of emigrants started going out of Independence 7 years ago and a lot of reports have come back down the trail. Celebrate well, folks. You got one hell of a ride ahead. Well, they got this far and look fit enough so they got a good chance.

A big man with a shiny face, probably the wagon master, blew the horn and we all roared a great cheer and went to bed. Mrs. O'Sullivan gave me a wool blanket. The blanket didn't much soften the hard wood of the wagon bed but nobody slept any better. I got a better blanket packed away on my Jackass but didn't get it out 'cause I wanted to be polite.

Me and the kids bundled up together in the back of the wagon. There's a bunch of sacks and boxes in the middle, and Mr. and Mrs. O'Sullivan sleep up front.

Mighty crowded but keeps us warm and the kids don't snore. Mrs. O'Sullivan has a little dainty snore.

The Captain blew his horn just before dawn. Maggie came over and pointed to a spot outside the wagon circle. Ya, I know what she wants. Guess she didn't wake up and sneak out while it was still dark. Well, neither did I.

I followed her out a ways and we took turns holding out our skirts. Yes, dear reader, there are not many outhouses out here on the plains. And yes, I could go into the embarrassments we women and girls have to endure with so many spying male eyes aching to see more, but breakfast is not far off so I won't.

And yes, I'm putting all these embarrassments right out here at the start of our journey together to warn you that this is not going to be one of those namby-pamby stories that petty everything up and lie up a storm.

So hello and welcome if you decide to keep on reading my story. So long and good luck with your boring life if you don't.

On my way back from the creek with a pail of water, a big, ugly fellow with a scar across his chin pulled me behind a wagon.

"I know who you are."

'You do?"

"Yep. You're a whore from Polly's place."

"No I'm not," looked scared and tried to get away.

"Ok little girl, let's make a deal. First chance we get, you're going to be real nice to me."

"What do you want me to do?

"You know what I want you to do. You just be real nice and I won't tell."

"Oh yes, yes. Thank you for not telling. Would you like a little kiss?"

He bent down and I hit that little bump on the front of his throat with the bony edge of my hand. He fell backwards, bug eyed and gagging.

"Mr. Ugly Man. If you ever tell anyone, ever, that I'm from Polly's I will kill you. Maybe I should have killed you here and now, but I'm a kindly person so I'll give you a chance. You understand me?"

People heard the gagging and came running. Captain Gray bent over.

"What happened, Ben?"

Mr. Ben saw me watching.

"I don't know, Captain. I just suddenly got a pain in my throat and couldn't breathe. I don't know what happened."

They carried the poor dear to his wagon. He vomited his guts out on the way.

"You stay away from Ben Carter," Mrs. O'Sullivan said as I handed her the pail of water. "He shouldn't even be in this wagon train. He's a lazy, no good bum from Tennessee"

I helped fix the beans, bacon and biscuits and we all sat around the fire and shared the food. Good Lord, whorehouse food is better than this. Don't they know how to fish? Or catch rabbits and prairie dogs with their bare hands? Ya, well, that's an old Indian trick and not many white folks know how to do that.

And I saw a bunch of pig nuts down by the creek when I was fetching water. She should have put some of those in the beans. Give them a little jump.

I watched the family as they ate. Mr. O'Sullivan is a small man with a light brown beard and long hair tied in the back. He seems in a whole world all by himself, dreaming of his farm in Oregon.

A whole 160 acres of rich bottomland in the Willamette Valley. That's a lot more than he ever had or ever would have in Ohio.

Mrs. O'Sullivan used to be pretty but is worn out. Like most of the women, she hated leaving her friends and the home she worked so hard to make comfortable and pretty, but Daniel insisted so what could she do.

The kids are lively. Julie is 7 and Daniel, Jr. is 9. They don't know if they're going to be jealous or glad to have me around. Right now they're glad because I told them an old Indian story last night about how two bunnies outfoxed a fox.

After breakfast all the wagons lined up. Maggie's is just back of mine. Then the bugle sounded and off we went with a big **Yeeeee! Haaaaa!** up and down the line.

It wasn't very lady like but I **Yeeeee Haaaaaed** right along with the best of them. Mrs. O'Sullivan frowned a little but I could tell she really, really wanted to **Yeeeee Haaaaa** right along with us.

Fort Childs is the first stop on the Oregon Trail, around 300 miles away. Army troops just started building the fort. We'll make around 15 miles a day so it should take around 20 if we don't tarry.

Well, except for the river crossings. The big ones will take at least a day to get across, maybe two.

2
THE UNDER WATER PEOPLE

So we're off to Oregon on the fourth day of May, 1847, in a whole bunch of dust, noise and bumps.

~~~~~~~~~~~~~~~~~~~~

Here is a picture I stuck in years later to show what it was like. They are way west of Independence and using horses to pull the wagons, not oxen.

~~~~~~~~~~~~~~~~~~~~

The seven wagons ahead of us kick up a hell of a cloud of dust. My hair and everything else have never been this dirty. And the wagon wheels squeak and squeal a terrible racket. Ya, they grease them. Makes them turn better but doesn't help much with the squeaks and squeals.

And the bumps! Oh my God, the bumps. They rattle your teeth and everything in the wagon. The fancy wagons have leather straps holding up the bed. Not this one.

There are a bunch of kids in the train, and two women well along with child. The men don't care. Hell bent for Oregon, wives be damned.

We got to Big Blue late in the afternoon of the second day and camped. Four Indians on beautiful ponies were spotted and everybody got a little nervous, but they're friendly. The Captain was told in Independence that the Indians help wagon trains cross rivers for clothing or a little money.

They gave the sign of greeting and friendship as they rode up, but the captain just stood there. I walked over and returned the sign.

One of the Indians signed the price for the crossing:
~~tomorrow~~cross~~4 shirts or 4 dollars~~

After a little bargaining the Indians agreed to 2 silver quarters each and supper. The Captain asked many questions about the trail and which Indians were hostile and which friendly.

After supper the Indians took me to a quiet place in back of a wagon. They know about the Blackfoot Spirit Woman who lives in the big house where women wide open their legs for pieces of paper.

And they know she was teaching the ways of a Spirit Woman to a little white girl. From the medicine bag hanging around my neck, and knowing sign, they figure that's me.

~~ you Spirit Woman?~~

I whipped out my Spirit Feather and smacked it across Black Wolf's nose. He grabbed his nose and sneezed, and the other Indians and laughed and laughed. I smacked two more noses and they just kept on laughing.

Then they quieted down and looked back and forth between me and the last Indian. I waited a good 2 minutes, and then wacked him a good one.

Good Lord, did they ever jump up and down and laugh. Like little children. I didn't know Indians laugh that hard. They are so serious and fierce around town.

Captain Gray came walking over. "Hey, what's going on?"

"Oh, they were telling me about two indian maidens falling into the river fighting over a strong, handsome brave."

"That sounds like a good story. Come on over to the fire and let's hear it."

Oops. Now what? Ok. They want a story. They get a story. Whorehouse style.

"Oh, I mustn't. It's a not a story a lady tells in front of gentlemen."

"We're all out here together in the wilderness, Mary. The old rules don't apply. We will think none the less of you."

I dropped my eyes and looked demure, like the girls in the parlor playing around with the fellows. We walked back to the fire in the middle of the camp and everybody urged me on.

"Well, ok."

I signed to the Indians:
~~me~~trickster coyote~~

"There was once a handsome brave, Two Arrows. He was strong and his arrows true. Two beautiful maidens fell in love with him and tried everything they knew to grab his heart.

"One night in desperation each in turn came to him and grabbed many other parts of his body. They did their best to show what a wonderful woman he would have.

"The next morning they sat quietly outside his lodge awaiting his choice. As Two Arrows lifted the flap of his lodge an old medicine man came over and pointed to the two women.

"You have made two mothers. You must take both."

"The two former maidens shrieked with rage. They did not want to share. They grabbed each other and began to fight. They rolled down the river bank into the cold water right below a waterfall.

"The Under Water People took them.

"Two Arrows was very sad and the next day went down to the river bank and pleaded with the Under Water People to give him back his wives. The Under Water People agreed but only if Two Arrows jumped into the water.

"The Under Water People took him too.

"The Under Water People like humans because they have long fingers and slippery tunnels. The Under Water People rub them together and eat the thick, white juice.

"Now when you walk near the waterfall, after the sun has gone away, listen quietly and stand very still. You will hear the wild cries of Two Arrows and the two women. Cries of sadness and joy. Wild cries of pleasure and pain.

But do not tarry too long. Do not tarry too long there on the river bank, listening to the sounds of the Dark Cave beneath the waterfall.

If you tarry too long, the Under Water People will take you too."

Black Feather signed:
~~I also fool whites~~know white talk~~good story~~ tell village~~from Coyote Woman~~strong Blackfoot Spirit Woman~~

Everybody was up early, spreading skirts, bringing in the stock, cooking breakfast.

Then began the hard work crossing Big Blue. We women helped unload the wagons and then the men took the wheels off and carefully slid the wagon beds down the steep slope to the river bank.

The Indians have slung a heavy rope across the river tied to a big tree on each bank. A rope was tied to the wagon bed and slung over the Indian rope and tied back to the wagon. That stops the wagon from floating down the river. Then a rope is tied to the front of the wagon, an Indian swims across, and the men haul away.

Most all the wagons made it across without any trouble but one started taking on water as soon as it went in. It was hauled out and recaulked, and the owner got himself a lousy reputation.

Another started to take water half way across and the men had to haul mighty fast and mighty hard to make it to the other side.

Then the wagon beds were pushed and pulled up the other bank and the wheels put back on. We filled Indian boats with our goods, and after many trips everything was brought to the other side and reloaded onto the wagons.

Yes, that work is as hard as it sounds. And Dangerous. One man almost drowned but an Indian saved his hide.

It took all day and half the next to get the wagons, goods, and livestock across. We stayed at Big Blue the rest of the day resting up and repairing wagons damaged in the crossing.

We made good time the next day and hit the junction early afternoon. The sign on the road to the south was marked **SANTA FE ROUTE** and the one to the west **ROAD TO OREGON**. Not much of a road.

We pushed on and made it to Observation Bluff late afternoon. Me and Maggie climbed to the top because we heard there was quite a view, and there surely is. You can see the wagon road going way off into the distance. They call it "a close look at infinity".

Osgar missed Aokii'aki and went back to the sporting life at Polly's Paradise.

The next day around noon I was riding Charlie next to Captain Gray when we came to the limestone cliffs overlooking the Wakarusa. Five Ponca Indians were waiting, and I signed greetings.

One asked me to tickle his nose with my spirit feather. News runs fast in the Indian world. He grabbed and sneezed, and they all danced around and around, laughing and laughing. They surely know how to have a good time.

Where are the vicious, bloodthirsty savages who scalp little children and shoot burning arrows into people they capture?

Captain Gray didn't know what was going on and looked confused and a little scared. But we got down to bargaining and the Indians settled for a silver half dollar apiece.

Oh My Lord! We thought Big Blue was bad. Before we could do the rope across the river trick, we had to use ropes to lower the wagon beds down the limestone cliff to the river bank. Then after they were safely across we had to haul them up the cliff on the other side. Two days of hard work.

The next few days were easy and we made good time. We began to see more and more graves by the side of the road. Most had been dug up, either by Indians for the clothes or by wolves and coyotes.

We had our first death. Dane, a 5-year old, fell off the back of his wagon and was trampled by the oxen in the next wagon.

We know the grave will be dug up soon enough but what can you do. Dane's father wrote out some words on a board and stuck it in the ground.

> Tis but a little grave, but O have care.
> Many hopes are buried here.
> How much love, how much joy,
> Is buried here, with this darling boy.

We had our first bad rain storm near Fremont Springs. The cold wind rocked the wagons back and forth and two wagons tipped over. The rain came down in buckets and there was no escape.

Ya, the canvas bonnets were treated with linseed oil but that only worked for a while, and then the water dripped through.

Hiding underneath the wagon bed just got you all muddy. So we just sat it out next to each other in the wagon, trying to keep warm.

After the rain stopped, it wasn't much better. Mud everywhere and nowhere to wash, and wagons got stuck and had to be double teamed.

Oh my good Lord, we have only gone around 200 miles and there are 2,000 more to go. Miles of rivers, mountains, and blazing hot, waterless deserts. This trail life is not for me. I don't want 160 acres of prime bottomland in Oregon. Well when I think about it, I don't know what I want.

People got more and more nervous and grouchy. Some saw the elephant and got crazy worried they made a terrible mistake. Fights broke out among the men, and women found little things to worry and bitch about.

Things were better when it wasn't raining and there was grazing for the stock, and wood for the fire.

At the Little Blue the trail ran between a high cliff and the river, barely wide enough for the wagons. Old Jake got hit by a big rattler. He was doing poorly anyway and died in the middle of the night.

For the last few days, I saw a dark cloud around him and two flashing knives of fire.

3
FORT CHILDS

A few days later, there, finally, off in the distance was Old Glory on a flagpole and around 50 soldiers and a couple dozen Negroes building buildings. The location is on high ground a little ways from the Platte.

Weariness slipped away and I got back my wild excitement exploring the Wild West.

~~~~~~~~~~~~~~~~~~~~

I found this old picture years later and put in here between the pages where it belongs. The fort wasn't this completed when we saw it. The flagpole is in the right place, but the two story building was only half built.

You can see the dust of the wagon train, and just a little of the Platte on the left, above the scrawny bushes

~~~~~~~~~~~~~~~~~~~~

Not much of a fort. No tall barricade to keep marauding Indians out, and the buildings were half built sorry affairs of rough cut sod and poorly made adobe brick.

Well, there are a couple of small, well build brush huts just west of the fort with a demure Indian maiden setting quietly in front. Hmmm. Wonder how they make their bread? Their ample dimensions suggest they have plenty of bread.

Lt. Daniel P. Woodbury came out to meet the wagon train and told us there was good camping just up the river. We're going to stay here today and tomorrow to rest up and let the stock graze.

The Platte is very wide but shallow and kinda sick looking but better than nothing so we women had a good bath thanks to our wide skirts.

Six Pawnee squaws came around late afternoon to barter. They wore tanned doeskin well decorated with trader beads and colored porcupine quills. They want cloth clothes, and anything metal, like knives, pans, kettles, bracelets and necklaces.

In exchange they have beautifully decorated leather moccasins, shirts and pants, and dried meat and berries, prairie turnips and pig nuts. I helped with the bartering and one of the soldiers noticed.

"You sign talk?"

"Yes I do," and he went back to the fort and came back a half hour later.

"Miss, may I talk to your parents."

"I have none. I'm an orphan, riding with the O'Sullivans thanks to their kindness."

"Ah. Then would you please come with me. Lt. Woodbury would like to talk to you."

The soldier led to what is going to be a two-story building. The bottom floor is pretty well finished. Lt. Woodbury arose from his desk as I entered. He's a proud man with full beard and a big pointy nose.

His roving twinkly eyes suggest he fancies himself quite a lady's man. Wonder what he'll do with a cute, innocent little girl.

"Thank you for coming, Miss. What is your name?"

"Mary Faraday, sir."

"Corporal Custer tells me that you sign talk. Is that correct?"

"Yes sir, it is."

"How well?"

"I was orphaned when 9 months old, sir, and raised by a wonderful Blackfoot woman. I have sign talked all my life."

"Excellent. Corporal Custer also tells me you have no kin in the train. Are you beholden to anyone in any way, like made a promise to stay with them to Oregon?"

"No sir. Mrs. And Mrs. O'Sullivan are kind people, but I have made no commitment to stay with the train."

"Excellent. Then I have a proposition to make. I very much need your services to communicate with the local Indians. I can offer you food, a small room of your own, and 15 dollars a month.

"Your job will be to grease communication with the Indians and go out occasionally and observe and visit the Indian villages nearby. If you serve well your pay will increase over time."

"Oh sir, that is so kind and generous. Thank you. I feel a strong obligation to ask the people who have taken me this far for their advice."

"Of course. I have heard that your train will leave day after tomorrow so there is a certain urgency for your answer."

"Thank you, sir. I will return soon."

Oh good Lord, what luck. I don't want to eat dust, but I do love adventure, and get so excited by the vastness all around me. I'll miss Mr. and Mrs. O'Sullivan and Sonny and Julie.

Maggie hugged and we cried our goodbyes.

Everybody waved good wishes as Charlie, me, and my jackass rode away.

Lt. Woodbury stood again as I came through the door.

"So. What have you decided?"

"Mr. and Mrs. O'Sullivan were concerned with so many men and no women. And they think $15. a month is a little low. They believe $25. more reasonable."

"Oh you are not the only woman here. There are three others and a number of negro women. As for your pay, I am sorry but am not authorized to offer $25."

I have had fun watching the bargaining at Polly's and have done a little myself for the treasures I have to offer. So looked sad and got up to leave. He motioned me to sit back down and made a show of going through some papers.

"Ah, here it is. Yes, if your knowledge of sign is as good as it appears, I can offer you $20. a month. I am sorry but can go no higher."

"May I meet the ladies before I make my decision?"

"Of course," and motioned to a soldier. "Go make sure that Betsy Ann, Delilah and Suzanna are ready for visitors. Tell them a very young lady would like to meet them."

Betsy Ann? Bottomless Betsy? She quit Polly's 5 or 6 months ago and disappeared. We had such good times together. Ya, now and then a particularly nasty drunk bastard had his pockets cleaned and rings yanked off his fat fingers.

And we experimented a little.

Lieutenant Woodbury led to another building with a mostly finished first floor, down the hallway to a kitchen. Yep, it's Bottomless Betsy.

She'll do a damn good job acting like she don't know me. A whore's got to be a good actress. You gotta act like you're getting plowed by the best farmer in the field.

"Miss. Faraday may I introduce you to Miss. Betty Ann Anderson? Miss. Anderson, Miss. Mary Faraday."

I smiled into her eyes as I took her hand. "Oh I am so glad to meet you, Miss. Anderson. Please call me Mary."

"Delighted to meet you too, Mary. Please call me Betsy."

Suzanna is the washerwoman and maid. She's tough and strong but not too bright. She has 12 negroes to help.

Delilah is a tiny scared mouse and is the fort's bean pusher. God knows what she's doing out here in this rough land. She has six negro women helping cook.

Betsy's job is to manage the two women, the eighteen female slaves, keep the books, and order supplies.

Later Betsy and I got together in her room, and hugged and kissed. "Oh Mary, I've missed you so. I didn't know how much I cared until I left and you were no longer near."

"I missed you too, Betsy. You were my best friend at Polly's, and we had such good times together. You respectable now?"

"Ya, I'm playing the straight life. It's boring as hell and surely don't pay much. I'm thinking of starting a nice clean house somewhere out west. Interested?"

"Nah! I had great fun playing the sweet, shy virgin but I'm getting a little old for that. Too much disease and abuse playing a regular girl. And I don't like being treated like dirt by those fancy bitches on the other side of town with their fancy dresses and powdered noses stuck in the air, and having to take care of those fat pigs with the badges. And I don't like drunks running their dirty hands over me."

"Oh Mary, you get used to that after a while. I can scream up a hell of a storm and plan what's for breakfast. Of course there were a few gentlemen I liked, and we had a good time together. Almost married one from a wagon train but turned him down because I was afraid of all those bloody Indians and all those deserts warmed by the fires of hell.

"May I ask what you're doing way out here so far from your clean little room in the basement of a house of fallen women?"

"Polly hooked me up with the wagon train that just came in. I sign good, so I'm the new army scout. Of course I'm not a real army scout. I'm just a girrrrrrl, but I'll do until the army sends one or one wanders by.

"The people here seem friendly. Any problems?"

"I assume that the good Lt. Woodbury failed to mention that the Missouri Mounted Volunteers, around 400 strong, are on their way from old Fort Kearny. The Mounted Volunteers will surely bring their own Indian agent who knows Pawnee and sign."

"You assume right, that slimy bastard. So I will soon be out of a job."

"No problem, Mary. We will just set you up in a fine establishment right next to the two Indian maidens. I promise you that you will have far more work than you can possibly oblige."

I made a face.

"Ok, Ok, then! I can see that that may not suit you. Then how about this as a fine alternative. When the troops arrive I will need an assistant and I believe you will be quite high on the list of applicants."

"What an honor it would be to work for a lady of such refinement, beauty and intelligence."

"Yes indeed, how lucky you will be. As for problems, right now the fort is well run and mostly safe but take care with Jake, that skinny soldier with the long black hair. He gets mean when he's drinking and horny as hell. Raped Delilah twice but she's too scared to report him.

"The other soldiers will try a feel now and but they back off easy when they see you are about to give them a good whack. The lieutenant does not seem interested. Maybe he's a Johnny man.

"Another warning for you. Watch out for Delilah. She has sticky fingers. Ya, she's a scared little rabbit but will steal you blind if you give her half a chance."

I got up early and waved as the wagons went by. That was one hell of a ride from Independence, and those poor folks have just started down the wagon road. I'll miss them but surely glad I'm not with them.

I cried a little as I watched the white bonnets slowly get smaller and smaller. Then disappear way off over the horizon. Into the vastness.

       Bye, bye, Maggie O'Hennessy.
       Fare thee well.

A couple of days later Charlie and I rode up the Platte to a Pawnee village. After a few miles we stopped on a little rise and I watched, and felt, and heard, the ocean of green grass, slowly waving back and forth in the gentle breeze. The earth glowed and trembled, and I heard the howls of the past and the roars of the future, and felt the deep vastness of wonder and delight all around me.

I sat there quietly on Charlie for a long time. Those were the happiest moments of my life.

The Pawnee village was maybe 30 earth lodges, a few brush huts, and a large lodge in the center. A woman just outside the village saw me and let out a yell.

There was a big hullabaloo and a half dozen warriors came out to meet me. They pointed to the medicine bag around my neck and let out loud shouts and grunts. Information travels fast in Indian country.

~~Coyote Woman~~come~~, and led to the big lodge.

A dozen or so warriors sat around the fire and four got up and sat down in a row across from me with their arms across their chests.

~~Make Spirit~~

I slapped the first three across the nose with my Spirit Feather and got the grab, sneeze, and howls of merriment. Then all watched the fourth man with great intensity. I waited a good long time and they all sat perfectly still. Waiting.

Then smacked him a good one. He grabbed and sneezed and everybody went wild. I have never heard such pure laughter and delight in white folks.

Well, maybe in very young children before they get civilized.

An old Indian, probably the chief, got up and raised his hand.

~~Tell story~~~Two Feathers~~~two maidens~~

There were loud grunts and shouts of appreciation when I finished signing the story.

Then all became quiet and a pipe passed. It was no-thought time. A quiet time each spends in their own way. I feel and know the plants, the animals, the earth, and the spirits all around me.

Jennie and I spent many hours in no-thought time touching the vastness inside ourselves and the vastness all around us.

Long Knife, the chief, got up and anger time began.

~~Great White Father promise~~ many goods~~land for fort~~send little~~

Warriors shouted agreement.

~~Great White Father promise~~white man no hunt north of village~~many white man hunt north of village~~winter hard~~

Another angry shout of agreement.

~~Great White Father promise~~~Sioux and Cheyenne no attack village~~Sioux and Cheyenne attack village~~burn corn~~steal horses~~children hunger~~

The shouts of rage were getting scarier and scarier. Yes, these are the murderous savages I heard so much about. And I was getting worried that if one of the warriors got angry enough he would push my face into the ground, run a circle around the top of my head with his sharp knife, and rip off my scalp. And walk proudly around the village with my long hair on his belt flapping in the breeze.

~~Great White Father promise~~no wagon cross sacred land~~ near mountain where ancestors sleep~~wagons cross sacred land~~

A great roar of rage shook the lodge.

~~Coyote Woman~~tell Soldier Chief~~

~~Long Knife~~I tell Soldier Chief~~

All became quiet once again, and the rage drained away. Squaws came with food and drink and we ate and drank. After a respectable time I signed thanks and goodbye.

After an hour wait I was ushered into Lt. Woodbury's office.

"Sorry for the wait, Mary. Would you like coffee?" An orderly quickly brought two mugs of coffee and large pieces of sweet cake on a wooden tray.

"So, what did you learn?"

"They received me with courtesy but their anger against the Great White Father is great. They accuse the Great White Father of not giving them the trade goods promised in exchange for land for the fort. They accuse the Great White Father of breaking the promise that there will be no hunting north of their village. They accuse the Great White Father of not protecting them from the Sioux and Cheyenne. And they are very, very angry that the Great White Father allows wagon trains to cross the sacred land where their ancestors sleep."

"Yes, that is what I have heard. Return to the village and tell Long Knife that a Head Soldier Chief will soon come to the fort and answer their concerns."

"Are they right?"

"That is not for me to determine."

"Do you have a copy of the treaty?"

"No."

"Why are you lying to me?"

"Because it is none of your concern."

"May I see the treaty?"

"No. Your job is to follow orders. And I order you to return to the village and tell Long Knife that he will soon have talk with Head Soldier Chief."

I stopped myself just in time from double flipping the famous Woodbury nose.

"I will send two mule loads of corn and flour with you."

And was dismissed. Ok, he's a decent fellow, just following orders.

After supper Betsy and I went for a little walk in the moonlight, and then went to her room and renewed our close friendship.

Afterwards she closed her eyes and talked about her greatest dream, a parlor house in San Francisco up on a hill overlooking the Golden Gate. She so wants to watch the clipper ships come and go. Oh, and a spiral staircase and French silk for the beds.

Ya, she knows that's a bit of a ways off and there might be one or two not quite so grand houses before she gets to sit on the long veranda of that grand parlor house on a hill.

And what is my dream? I don't have any. I guess I should but I don't. I just want food in my mouth and not have to be too cold or too hot or too wet, or too dry, and maybe a little rub and poke now and then.

But more than anything I want to explore the vastness. Yeah, I guess that's my dream.

I heard whimpering and whining as I passed Delilah's room. She's getting a good plow from somebody, or really knows how to do herself. I went out and around to her window. It was cloth covered except for a tiny hole in a corner.

Yep. Jake is giving her a mighty fine plowing. Nope, he isn't raping her. She's pushing back hard to get him in deeper.

4
RISING WOLF
An old trapper

A few days later I saw an old man on a painted pony ride into the fort, followed by a long string of loaded pack mules. Buckskin, fancy beaded moccasins, and a raccoon cap with a long beautiful black and white tail.

Clean shaven. I thought all those old mountain men and trappers grew beards and now and then chopped them short with their axe.

One of the negroes dropped his hammer and came running.

"Mr. Monroe! Mr. Monroe!"

Mr. Monroe jumped off his horse and they had a hell of a big bear hug.

"Oh my God, Sammy, I thought you were dead for sure. How'd you ever get out of that ambush?"

"Ran like hell, Mr. Monroe. And how did you get out of town with that Sheriff and his whole parcel of deputies chasing after you?"

"Same as you, Sammy. Ran like hell."

And they had one hell of a knee slapping, belly shaking laugh.

"Mr. Monroe, I need your help. They swear I'm another man, a runaway. Well you know that was no mistake, just a way to get a cheap slave."

Lt. Woodbury came out of his office to see what all the hullabaloo was about.

"Hugh! Hugh Monroe! You mangy varmint. What you doing this far south?"

"Down to Council Bluffs with some mighty fine peltry."

"You come all the way down from Blackfoot country? How'd you get past all them Sioux and Crow.

"I've lived a good while among the Indians and still have all my hair, so they changed my name from Rising Wolf to Ghost Rider. Strong medicine if you see or touch me. Very bad medicine if you harm me.

"Mighty fine to see you again, Lieutenant. Seems like the last time was Hell's Gate up the Missouri. You were building a road."

"Ya, you got that right. Never did get that road built. Bunch of Flatheads chased us off."

"Lieutenant, I believe there has been a mistake here. I've know this man a good long while. We trapped together many a season up Blackfoot country. His name is Sammy and he is a free negro."

"Well, come on into my office and let's have a look at the papers."

A couple of days later I was in the mess hall at noon and saw Mr. Monroe sitting at a table eating his biscuit and beans. He looked up and dropped his spoon.

"Miss. Could you come here a moment?"

I walked over and he stared at my chest. No, not there, a little higher, at the crucifix and medicine bag.

"Where did you get that medicine bag?" There were tears in his eyes.

"A Blackfoot woman who raised me gave it to me."

"And her name is Aokii'aki."

"Yes! Yes! How did you know?"

"Please sit down. May I see it?"

I took it off and handed it to him. He wiped tears more than once as he stroked and caressed the fur. "Aokii'aki. Aokii'aki."

"She was my woman long ago, in my trapping days. We spent many a happy season camped along the Two Medicine Lodge and Swift Water. I loved her more truly than any woman I have ever known.

"When I was in Montreal one summer the head medicine man convinced the tribe she was full of evil spirits and ran her off. I searched and searched but never found her. Is she still alive?"

"Oh yes, she lives in Independence." He gently kissed the bundle and handed it back.

"Where in Independence?"

"Where?"

I hesitated, and he jumped up and leaned over the table.

"Why won't you tell me!?"

"Please, Mr. Monroe. Can we find a place where we can be alone later today?" He didn't much like the idea, but sat back down.

"Let's meet after supper and take a walk."

After supper we went a little ways from the fort and sat down on a couple of rocks.

"You are an honorable man, Mr. Monroe. I must ask you to keep to yourself what I am about to tell you."

"You have my word, Mary."

"Jennie, your Aokii'aki, became my mother when I was 9 months old. We lived in a little room in the basement of Polly's Paradise, a whorehouse. Polly's Paradise is a high end whorehouse, the only one in Independence. It is a good clean house, up from low end houses, cottages, cribs and streetwalkers. Polly treats her girls and customers well.

"I do not know your attitude toward the sisterhood but hope you will not be too harsh or too quick to judgment."

He sat quietly, staring at the medicine bag for a good while.

"Was Aokii'aki mistreated in any way?"

"Never. Jennie and Polly are close friends and powerful women. No one would dare mistreat them."

"That is all that matters to me," and wiped another tear away. "I'm going to stay here a few days to feed and rest up the mules and then I'm off to Council Bluffs to deliver the peltry. Won't be there long and then coming back here.

"The main contingent of the Missouri Volunteers will arrive by then and I have some business with them. Then off to Independence. I will be glad to take you with me if you want to go back."

"Thank you. I may accept your kind offer. Can you keep it open until you return from Council Bluffs?"

"Of course."

Mr. Munroe and I ate lunch together a few times before he left and he told me a lot about himself, Aokii'aki, the early days of the Hudson's Bay Company, and the Blackfoot.

"I was apprenticed to the Company on the third day of May, 1814. My Grandfather was part owner. I was only 16 but quick to learn, and the Company badly needed someone who knew the Blackfoot language and their ways. So I was sent back with the Blackfoot when they finished bartering their peltry at Mountain Fort.

"Lone Walker, the head Chief, agreed to look after me and teach me their ways and language.

"One day on the way back, we stopped for lunch on a sunny day next to a small lake. After we ate, we sat in a circle and a medicine man took out his pipe and tobacco. After packing the bowl he opened his fire pouch but the smoldering spark had gone out. There were many grunts and groans of dismay.

"I jumped up and yelled 'I will light it for you,' forgetting they knew no English. They all turned.

"I grabbed the sun glass my grandfather gave me and motioned the medicine man to put the pipe to his mouth. First there was a little smoke and then a flash of flame. The medicine man took one long puff and then with a great shout jumped up and held the pipe high, pointing to the sun.

"It was very quiet, and then with a great roar they jumped up and rushed toward me. I knew Blackfoot worship the sun and figured I just got caught stealing the sacred fire. I figured my time had come.

"Nope. They crowded around with big smiles, rubbing their hands against me. When others heard they all came running. Squaws shoved warriors aside to rub their babies.

"After a time all became quiet and the leaders of the tribe made a place for me in the circle and the pipe was passed around. I did my very best not to cough when it was my turn, but I got a little sick.

"I brought down fire from the Sun so they gave me the name, Sun Fire.

"A couple of days later I was out hunting with Red fox, Lone Walkers' son. and came over a rise There in front of me was the vast plains stretching out forever and ever, and high mountains covered with snow to the south west. I was the first white man to see the vastness. The first to explore. The first to move freely among the Blackfoot and learn their language and ways.

"I was swept away by an overwhelming feeling of excitement and awe. That was the happiest moment of my life."

I clapped and laughed and told him I had the same moment. The moment when the earth glowed and trembled, and I heard the howls of the past and the roars of the future.

We share a deep understanding and bond. We know the power of the vastness.

"How did you get the name Rising Wolf?"

"They changed it after I saved Red Fox from a grizzly bear."

A couple of days later, Lt. Woodbury sent for me. He did not offer coffee or sweet cakes.

"Mary, 9 horses and 12 head of cattle are missing. There are signs the Pawnee are responsible. I would like you to scout the village and find if the missing horses are in their herd. All of the missing horses have the cavalry brand which you probably have noticed is **US** on the left shoulder.

"Don't let them know why you are there and return as soon as you can."

"I don't like sneaking. I would like to ride into the village and ask if they are responsible for the stolen horses and cattle."

"You have been given your orders. I expect you to obey them."

"No. I won't sneak and lie. Find someone else." And got up to leave.

His face began to change to a pretty pink, but then he took a deep breath. Impressive control.

"Mary, please sit down", and called an orderly to bring coffee and sweet cakes.

"Mary, there are two kinds of morality when dealing with your fellow men. There is the morality of individuals with each other, and the morality of those responsible for the lives of many. I am trying very hard to think of the lives of many, and that is why I am asking you to sneak and lie.

"If you told them the truth, you might well trigger a war and be responsible for many deaths on both sides. The odds of a war is reduced if you sneak and lie. Do you understand what I am saying?"

The orderly brought a tray with two mugs of coffee and sweet cakes.

"Yes, I understand what you are saying. I don't like it but I understand and agree there are times a leader has to lie to protect his people, and a Lieutenant to protect his troops. And I'm no Snow White. I've done my share of lying and I'm a very good liar.

"But the Pawnee trust me. If I lie, I betray that trust, and that I will not do, and when they find out I lied, I'm worthless to you."

He took a few sips of his coffee and looked back and forth between me and out the window.

"Alright, Mary. We'll go honesty. How about this? Carefully circle the herd and see if any of the stolen horses are there. I have a spy glass you can use. If they are not there go to the village and have a pleasant visit.

"If the horses are there, meet with the chiefs, say that we have horses and cattle missing and ask if they have any information where they might be. Let us see how honest they are."

"Fair enough. I'll go early tomorrow morning."

Yes, the horses with **US** on the left shoulder are there. I made a wide circle back to the trail. The Chief and two warriors rode out to meet me, but no cheering from the village.

The stares were not hostile, more watchful concern. We sat in a circle in the main lodge.

~~Why you come?~~

~~Soldier Chief ask~~you take horses?~~you take cattle?~~

~~Great White Father promise horses~~promise cattle~~no give~~we take~~

~~I understand~~wish peace~~fear war~~

When I left the chief's lodge the mood had changed from watchful concern to excitement and friendliness. The women and children surrounded me, touching, shrieking and laughing.

A pretty little girl came up and gave me a flower. She had long, beautiful, black hair hanging down her back, and wore a beaded and quilled doeskin dress. I gave her a big hug and a shiny penny.

Did she ever whoop and holler, holding up the penny for all to see. A woman nearby signed her name~~Dancing Deer~~.

Then rode back and reported to Lt. Woodbury.

"The problem Mary is that the treaty they signed requires them to move North of the Platte before we fulfill our end of the bargain. They refuse to do so. The Missouri Mounted Volunteers will arrive soon. Perhaps then a decision will be made what to do."

The day before he left for Council Bluffs, Mr. Munroe and I sat together for lunch and he promised to share the strangest experience of his life.

"Akishi and I were on our horses by the shore of upper Two Medicine Lake. He is a good friend and a powerful Blackfoot medicine man. He asked many questions about the white man's world, and worked hard to learn our language. He learns fast, and we were soon only talking English when alone.

~~~~~~~~~~~~~~~~~~

Years later I found a picture of Akishi and placed it here where it belongs among the pages.

~~~~~~~~~~~~~~~~~~~~

"As we sat there on our horses looking across the lake at the high, snow covered peaks, I suddenly felt watched. Akishi motioned me to be silent and not move. The feeling grew stronger and suddenly I saw a young woman standing behind us.

"I didn't see her with my eyes. I saw her with my mind. She was dressed very strangely and wore pants like a man, and a man's shirt that clearly showed her body. She wore high, shiny boots, black as tar. After a minute or so she faded away from my mind.

"Akishi explained the woman was from a time to come, when women dressed with so little modesty. A time very different from our own. A time when the Blackfoot live like dogs on a small piece of what was once their land."

Late in the afternoon, two days after Hughie and Sammy left for Council Bluffs, the 400 strong Missouri Mounted Volunteers marched proudly into the fort led by fife and drum. It was a grand sight and sound, and everyone waved and cheered.

Well, the cheering ended soon enough for Betsy Ann, Lt. Woodbury and me.

The Lieutenant lost his office and quarters to Lieut. Colonel Ludwell E. Powell. I hope the Colonel doesn't get too comfortable. His enlistment and the enlistment of the men in the Missouri Mounted Volunteers will soon expire, and in the early fall will be replaced by regular army units under the command of Capt. Charles F. Ruff.

Betsy Ann became assistant to the battalion quartermaster, and I lost my job to the Indian agent. Col. Powell graciously allowed me to remain in my tiny room, at least for a while. The fort's name was changed from Fort Childs to Fort Kearny.

Betsy Ann didn't get along too well with the quartermaster. It was a long drop from being in charge to being under the thumb of a dumb, authoritarian bastard.

We decided to return to Independence when Mr. Munroe comes back. He's the best chance I got to really see the vastness. I don't think he's going to spend the rest of his days rocking away on the front porch, whittling cute little puppy dogs for the neighborhood kids.

He'll return to Blackfoot country and Aokii'aki will go with him.

I can't wait to see Jennie's face when Hughie comes walking through the door. She sometimes talked about camping along the Swift Water with her man, but never mentioned his name.

Wonder if she shared? Blackfoot warriors often have more than one wife. The poor dears have to step in to satisfy the needs of all those horny maidens made surplus by their men getting killed chasing buffalo and battling the Crow and Sioux.

Following in the shadows of the battalion from old Fort Kearny was the usual grog shops, grocery stores, drifters and grifters.

And now the two Indian maidens face fierce competition. Betsy Ann and I watched with a professional eye as the newly arrived ladies set up their establishment.

Soldiers get paid poorly and camp followers working the sheets are not much above street walkers. Their tents needed washing and mending, and so did they.

Bottomless Betsey can't wait to have a house of her own. She doesn't have a lot of money so will have to work her way up to that grand parlor house on the hill.

A few days later a scout came riding hard, jumped off his horse, and ran into Col. Powell's office. A bugle sounded and the troops quickly caught and saddled their horses.

Sioux had attacked the Pawnee village, and killed and scalped many. Those who escaped are on their way to the fort. Two company of troops rode out to meet them.

A few hours later the troops returned with the Indians. There was blood everywhere. Some from Sioux knives , some from their own as they slashed themselves in grief and despair. We rushed to help them into the fort.

A young woman with a little girl in her arms came up to me and lay her at my feet. A shiny penny hung from her neck. She was scalped and blood still oozed from the top of her head.

The next day the troopers accompanied the Indians back to their village. One of the women asked me to go with her. We saw the smoke a few miles away. The Sioux burned the village and broke into many hidey holes.

The Pawnee warriors scattered because of the overwhelming odds and sent runners to a couple of nearby villages. The warriors have assembled and will leave early morning.

Yes, and if they win the battle they will do exactly the same to the Sioux village that the Sioux did to theirs. Of course the women will be left behind to clean up the mess and try to make a home for the warriors when they return, if they return.

Ah, the glory of battle. Sometimes I don't like men very much.

Hughie and Sammy came back from Council Bluffs the next week and were glad to have Betsy and me come with them to Independence. We left early the next morning.

I didn't look back. The scene of the blood, the pain, the despair, and Dancing Deer lying on the ground, will be in my memories and bad dreams for a long time.

That vastness I care so much about is just pumped up fantasies, lovely opium dreams.

Not to despair. My lovely fantasies will flood back soon enough. We humans cannot bear much Reality.

It only took nine days of easy riding to get back to Independence. We left our horses, mules and my jackass at the Independence City Corral. I just love it when the wind whiffs the sweet aroma to all those sensitive, high-born, refined ladies and gentlemen on the other side of town.

Was Polly ever surprised and glad to see us. Lots of laughter, joy, hugs, and kisses. She knows all about Hugh Monroe, I didn't know he was so famous, but she didn't know Jennie was his woman.

Oh I so wanted to be there when Hughie and Jennie meet, but he wanted to go alone. Can't blame him for that.

We waited, and waited and waited but no Jennie, no Hughie. We now and then went down to the basement. The door to Jennie's room stayed shut.

Well, they finally came into the dining hall around noon the next day. I'm sure you have all heard about the glow women have when they carry new life. Jennie and Hughie both glowed, but I don't think either is carrying new life. They merged together into the vastness.

"Did you two have a good night's sleep," Betsy asked with a smirk.

They both laughed.

"Why yes we did," Hughie said," but that was after Aokii'aki reminded me how a Blackfoot woman pleases her man."

Aokii'aki blushed. "And I go home. Long good friend Akishi now head medicine man. Start when sun melts snow. Come with us, Sopo'aki."

"Sopo'aki will come with you. I want to sit in the spirit circle on Two Medicine Lodge and Swift Water, and talk to the demons in the Lake of Many Devils, and watch as the great head of their mother slowly rises from the water."

"It will be so, Sopo'aki. Places of great power, many spirits. My home. Your home."

Later that day Polly, Betsy, and I met on the veranda of Polly's Paradise. and shared extra spicy ginger cookies and tea.

"So, what are you two young ladies going to do with your lives?"

"I'm going to Blackfoot country with Jennie and Hughie, to be a Spirit Woman and explore the vastness."

How about you, Betsy?

"I don't know what I'm going to do. I hope I can come back here."

"Of course you can back here, you were one of my very best girls."

"But you know what I would really, really like to do? Have my own house. I was in charge of people at the fort and took care of the books and supplies, so I know I can do it."

"Ah, so you want to drive me out of business. Some gratitude."

"Of course. All I need is enough money and a little help how to start a high end house."

"Actually there is plenty of room for us both, and if you don't start another high-end house in Independence, someone else soon will. In fact, there is a quite suitable building the other side of the Independence Corral, and I hear the wind seldom blows in that direction.

"We could trade girls when the fellows get tired of the regulars, and could cut costs by ordering supplies together. Two ladies of the night, sharing the good life."

"Just one small problem. I may not have enough money. How much do you think it would take?"

"To buy the house, bring it up to high end standards and get the girls and get them ready? Probably around $5,000."

"Ya, I was afraid of that. I've only saved around $1,500. Oh well, I'll just keep saving, and maybe someday."

"Betsy, if you make me a fifty-fifty partner and manage the house I'll supply the rest."

They both looked so shocked, and then Polly slapped her fist up the side of her head.

"You're the one who grabbed all that money from the banker."

"Oh my goodness no, Miss. Polly, old friend. I would never do anything like that. I have just been very, very thrifty, and saved and saved. You are aware I am sure that I entertained a few nice gentlemen who wanted so bad to guide a poor little girl into the joys of womanhood. They paid real good."

"Oh yes, I'm well aware of all that, and I am also well aware there in no way in hell you could have saved that much. But wherever, whatever, however. You really have $3,500?"

"Yes, I do."

Betsy Ann just stared at me for the longest time and then came over and gave me a hug. There were tears in her eyes.

"You got a deal. You put up the $3,500. and I manage the house and share the profits, fifty-fifty."

"I'll help all I can this winter but you know I'm off to Blackfoot country in the spring."

"Your half of the profits will be deposited in the best bank in town. A deal?"

"A deal."

"Ok, you two enterprising young business ladies, and future madams of the second best whorehouse in Independence, come on over to my office tomorrow morning around 10 and we'll start planning."

5
THE SECOND BEST WHOREHOUSE IN TOWN

"Ok, ladies. The first step is go see old Doc Davies to buy the house. He wants $2,350. Ok if I do the bargaining? I know him well and am pretty good at lowering a price. And of course raising a price."

Chuckles all around.

Doc Davies is pretty much retired and was home. Offered coffee, straight or Irish, exchanged a few pleasantries, and got down to business.

"I assume you ladies are here to talk about buying the beautiful home I have on the market."

"Why yes, Dr. Davies, if you are referring to that run down, beat up shack you have for sale."

He laughed. "Ah, Miss. Polly, I see that this will be a long, enjoyable bargaining session. What are you offering?"

"What are you asking? I believe the seller sets the price."

"I have recently lowered the price and it can now be had for the outrageously low price of $2,500."

"Oh my, Doctor Davies. We are both fully aware why you lowered the price. It has been on the market for quite some time and there are no buyers. And I do believe you have made a small mistake in the price. The last price you posted was $2,350."

"Oh Miss. Polly, I am so sorry. You are so right. I am getting up there in age you know. My memory has never been very good, and I fear is not getting any better. What is your offer?"

"I am ready to buy your house for a very generous $1,500."

"You are aware of course that is a very low offer, and if I didn't know you better I would assume you were attempting to insult me."

"Oh Doctor Davies, I would never ever do that."

"I know. I know. Just funning a little. We have known each other for a long time so I am willing to be very generous and will hand over my precious home to you for $2,100."

"That appears to me a rather tiny bit of generosity, Doctor Davies. I can go $1,700. and no more. Take it or leave it." And got up.

"That is hard, Miss. Polly. The house is worth much more. How about an even $2,000?"

"Oh you are a hard man, Doctor Davies. I will raise my price to $1,850." And moved toward the door.

"I am temped but I cannot. I will just have to wait a little longer to find a buyer."

"What if I throw in a free girl once a month for a year?"

"You throw in one of your best once a week and we have a deal."

"Done! Thank you, Doctor Davies. As you are well aware, my girls do not disappoint."

We spent the rest of the pleasant visit, sipping Irish coffee and munching assorted bunny cakes and talking local and national politics and the price of onions.

"What a great job, Polly," Betsy said, on our way home. "Thanks."

"Sure. Nothing like dangling a bunch of beautiful, willing women in front of a man. They turn into little hungry boys reaching for the cookie jar.

"Now the hard part. Refurbishing and recruiting. I have a man I've worked with for a long time. He knows what he's doing and is reasonably honest and fair. Mr. Jacob Allison. Do either of you know him?"

I laughed. "Oh my yes, he made me a woman on a mattress in the basement while his wife was having tea and crumpets upstairs in the drawing room with her lady friends. He was gentle and generous."

"Ok. We'll close the deal for the house day after tomorrow and Jacob can start work the next day. The first priority is a well appointed parlor with a good, reasonably priced bar where our guests can make their selection.

"Some houses try to make more money from the bar than the girls. You try that and a lot of fellows don't come back."

We ended the evening gossiping about the crooked mayor and who got caught sneaking out of Miss. Abigail's room. We met Polly the next morning to continue planning.

"Fall and winter are slow time, so we have a few months to get ready. As we refurbish and gather the girls we can offer a few very lucky men a very special preview of the ladies, at a very special price. And then just as the wagons start to roll, we'll have a grand opening,"

"Must be real hard getting enough girls."

"No, not really. Betsy and I know girls in the two low-end houses, and the cottages and cribs. Soon the girls will be coming to us, begging to get into a high-end house."

"Won't that start a war?"

"Maybe. I fought a few battles in my day so most of the business ladies in town know not to mess with me. If they try, I have some friends who will come visiting."

"I've always felt sorry for the poor street walkers. Let me spend a few nights looking for girls who might be suitable."

"Sure Mary, give it a try but don't get your hopes up. And be careful. There are mean bastards in the shadows.

"Ok, let me tell you what kind of girls I think best. Here are the big four: they like sex; they have a genuine, pleasant get-along personality; have at least a little experience in the trade; and are clean. And by clean I mean that they don't have a disease and that they are careful to keep themselves well washed.

"Some of the 15-minute girls in the cribs don't even bother wiping themselves.

"Beauty is over rated. Sure it's nice to have but you won't get much repeat business if you don't have at least most of the big four.

"A few guys come just for sex, but most come because they miss the companionship of a woman, someone to hug and pat them like their mother used to do. And many are sad, lonely and home sick, especially guys off the wagon trains. We give them what they need.

"Oh! And be sure and hire a couple bouncers. You don't want them to be mean, just look mean and able to do the job if a drunk starts trouble. They come cheap for the benefits."

All I saw the first night wandering the streets were poor, sad girls way past their best days. They probably worked their way down the ladder from houses to cottages to cribs to the street.

Most are high on laudanum or whiskey. Maybe I'll end the same, begging for food or just a little sip of that bottle, offering to do anything. Anything!

I told a couple of women who looked healthy enough that they should hop a wagon train and try their luck out west. Plenty of lonely guys aching for the comfort of a warm, loving wife and not too particular about their past, age or beauty.

The second night I saw a possibility. She was young, clean, and didn't look high on anything.

"Hi, I'm also a working girl. How's business?"

She looked me up and down.

"God damn it, you're one of those holier than thou bible thumpers, out to save me from my devil ways. Get the fuck out of my face."

"Great God almighty, how you do swear. What put a burr up your ass? I just want a little friendly talk."

She looked me over a little more carefully.

"Damn, you really fooled me with that sweet, innocent face. Probably working the virgin game like I did long ago."

"You new in town, or get dumped?"

"I'm new. Got run out of St. Louis by those damn bible thumpers. That's why I went off on you like that. Stupid, hypocritical bitches. You walking or got a place?"

"Me and a friend are starting a high end house and could use your help setting it up. We can give room and board and $25. a month. Not much, but if you work out you'll get a nice room upstairs."

"You don't look old enough to start a house. You for real?"

"Well I fooled you once. Looks like I fooled you twice." She laughed.

"Here's the address. If you're interested come around for lunch tomorrow. Nothing fancy but will put a little fat on those skinny bones. My friends name is Betsy Ann, and I'm Mary."

"Very pleased to meet you, Mary. Mine's Harriet. See you tomorrow."

And then there was Tabatha.

I heard cries of fear and pain coming from a dark alley. A big guy was brutally humping a woman pinned against the wall. She was desperately trying to get away but he was way too strong.

I kicked up between his legs with the flat of my foot, hard. Jennie assured me that pretty much destroys a man, at least for a while.

She was right. He went down howling and I don't think he'll be getting back up for a good while, and for sure won't be raping anybody any time soon.

The girl was wild eyed scared and grabbed me. I adjusted her clothes and led her out of the alley.

The bastard was still on the ground, puking his guts. Poor, poor dear.

I took her a few blocks to a small eatery and after a while she calmed down enough to sip the tea.

"What happened?" She looked up with big, teary eyes.

"I was just walking down the street, and he grabbed me and pushed me into the alley. Then he pushed me against the wall and grabbed my dress!" And started to cry again.

"There, there, you are safe now." She took another sip of tea. "What were you doing out alone so late at night?"

She looked confused, and lowered her eyes as she took another sip of tea.

"What were you doing out alone, so late at night?"

"I wasn't doing nothing. I was just walking around."

"There is only one kind of woman who walks alone in that neighborhood late at night."

She looked so desperate and started to cry again. "I'm not a whore. Don't call me that. I'm not a whore!"

"I just saved you from that bastard. Be straight with me."

She looked away and then back. "My husband doesn't want me anymore and left on the wagon train that pulled out two days ago. I don't know what to do. I have nowhere to go."

You know how to sew? Cook? Clean?"

I am a very good seamstress, not so good a cook but I get by. I clean if I have to. Why do you ask?"

"Here is an address. If you want a free meal and maybe a job and place to stay come on over tomorrow noon."

"Thank you. My name is Tabatha Parker."

"Hi. I'm Mary."

Altogether I gathered 7 lost souls. Some are working girls and some will help get the house ready, and keep it neat and clean when we open. They all have a story, some tragic and sad, some full of fun and adventure. Too many stories to write them all down.

Oh well, one more. The three nuns.

It was early Sunday evening and the church bells were ringing, and people were hurrying for late mass. I got a note from Sister Mary Ellen to meet her outside the church. Real curious what she wants. Never talked to a nun before. Not many hang around Polly's Paradise.

I watched the people crowding through the church door for a while, and then a nun in the usual black and white came over.

"Are you Mary?"

"I am."

"Please come with me."

She led around to the back of the church, into a stone building. Two other nuns were sitting around a small round table.

"This is Sister Bethany, Sister Sophia, and my name is Sister Mary Ellen. We would like to talk to you if you have a little time available."

"Of course." Sister Bethany served tea and berry tarts.

"Thank you for coming, Mary. We are aware that you are a member of our other sisterhood and Polly's friend. We work with the sick, the dying, the lonely, and the despondent, and know that Polly's Paradise is not quite the Devil's playground that our hypocritical priests thunder against from the pulpit.

"In talking to the men who have frequented Polly's Paradise we are aware that the girls at Polly's are more than just bodies, but are often quite understanding and kindly. Is that true?"

"Yes, that is true. There are houses that are the Devil's playground just like the priests say, with dishonest, uncaring and greedy madams and girls. Polly's Paradise is not one of them."

"We have also heard that you are looking for girls to populate a new establishment."

"You have excellent spies, Sister Mary Ellen. That is correct. We are opening a second high end house in Independence."

"We three would like to be considered. The two priests here have made us their whores. We have decided we would like to be paid for being so."

They were serious. Obviously nervous but their jaws were set and there was fire in their eyes.

"I can see that you are serious, but perhaps you are considering this change of profession a bit too hastily."

"Our decision has not been made hastily." The other two nuns nodded

"Ok, I admire your courage, but not sure of your judgment. I can think of a big problem right away. If you come over to the other sisterhood there will be hell to pay when people find out. There will be a huge hullabaloo and the mob will run us all out of town"

"Wait here." And the three nuns left. A few minutes later they returned, wearing beautiful, fashionable clothes that showed off quite lovely bodies. Their hair was well kept and full and they did something to their faces. Painted faint lines maybe?

"Do we pass?"

"Yes, you pass. Here is an address. Come to lunch tomorrow the way you are now. We will decide if we want to take the risk."

######################

I have not been able to read the rest of Mary's story. The remaining three-quarters of the pages are written in pencil and most too faded to read. I am aware there is technology available that may be able recover some of the rest. If successful, I will add them.

######################

######################

I have found a company in town that has an array of special lights that can often illuminate old manuscripts. So far they have been able to recover the next five chapters. Not all parts of the chapters were recoverable and I have had to interpolate and extrapolate a few times.

For instance, I am not sure the paddlewheel steamship going down the Sacramento River was the El Dorado, nor whether there were two or three smiling ladies in the early days at Dry Diggings. Some of the dates and mileage may also be off a bit.

######################

6
CALIFORNIA GOLD

A big shock hit Independence in the spring of 1848. **GOLD!** Big nuggets of **GOLD** just lying there on the ground. Just waiting for you to bend over and drop them into your honey bag.

I never saw Osgar, that little, horny Irish Elf, so excited. He jumped up and down shouting "**It's true!! It's true.!!**" urging Mr. Munroe, me and Aokii'aki to get quick to California.

Gotta get our pot of gold before the whole country catches gold fever and comes riding across the plains, sailing around the Horn, and running across the Isthmus.

People back in the states are still recovering from bad times, so not much to lose racing for some of all that California gold.

Osgar has assured me that he can smell gold a mile away. Mr. Munroe knows about Osgar but not his nose. Our friends have no idea he's around, which is just as well. He spies on the women ever chance he gets.

Mr. Monroe and I asked Betsey to come for a little meeting one morning.

"Mary and I are putting up the money for a trip to California and Sammy is bringing 6 negro friends, all strong men, to do the heavy lifting. We'll act like the negroes are ours but they are all free negroes, free to come and go as they please.

"Betsy, would you or any of your girls like to join us? We sorely need a few women around the camp to add a little comfort and good cheer, and I imagine there will be plenty of opportunity for enterprising women in the mining camps."

"I been thinking a lot about all that gold, Mr. Monroe. I figure this is my big chance for that Parlor house on the hill. Yes, I'm in. I've already talked to Polly and she'll take on the management of our house, and split the profits three ways with me and Mary.

"And I'm pretty sure some of my girls will want to come. This is their big chance to start a whole new life and maybe grab some of that gold. I'll bet the three nuns will be at the top of the list."

Yes indeed, the three nuns surely are at the top of the list. They are having one hell of a ride. Raised in good Catholic homes, took the vows, drilled good by two holy priests, joined the other sisterhood, and now riding across the plains and mountains to get their share of California gold.

Yeeeee! Haaaaa! What a ride.

We're going to be quite a bunch riding over the high plains and through the mountains. Coming along with Bottomless Betsy are Molly, Harriet and the three nuns. Harriet is that smart, tough girl I recruited who got kicked out of St. Louis. And then there's Aokii'aki, Mr. Monroe, me, and Sammy and his six friends.

Two of Sammy's friends are blacksmiths and the other four are their helpers. They have prospered and saved, and have more than enough to buy mules and provisions to make the trip to the gold fields.

They're working night and day turning out gold pans, pots, pickaxes and shovels. Just the iron part of the pickaxes and shovels. The miners can cut and carve the wooden handles.

We are going to raise more than a few eyebrows with seven negro men traveling along with seven pretty, young white women. Got to make sure they are far apart whenever we hit a settlement or meet people on the trail.

When we reach the gold fields the negroes will continue to act as if they belong to Mr. Munroe to better avoid people taking advantage of them or causing nasty confrontations.

Are we all going to get filthy rich or are we all just chasing a dead cow? Oh well, if the gold is a bust, we're off to San Francisco to work our way up to that grand parlor house on the hill, watching the graceful clipper ships slip and slide through the Golden Gate.

It'll be a damn good ride whatever happens.

Mr. Munroe knows the trail pretty good from his travels and talks around the campfire. We got to hurry before the Sierra passes get snowed in, so we all worked fast and hard to get ready.

No wagons, just horses and pack mules. We'll travel much faster.

The blacks are bringing 16 mules to carry provisions and the iron goods, and 9 mules to carry needed items for the blacksmith trade like anvil, forge, bellows, hammers, tongs, iron bars, etc. They'll make their own charcoal when we get there.

Hughie and I are bringing 35 mules packed with trade goods. He has bought just about every pair of boots for sale in Independence, along with much of the tobacco, lead, powder, and numerous other items badly needed in a mining camp.

We have 15 mules to carry our provisions, tents, water kegs, personal stuff, etc. We'll hunt and fish along the way. Oh. And a large white tent for our first mercantile emporium.

We kicked up quite a storm as we rode out of town. The gold bug bit lots of folks but it's too late for most to be starting out. They didn't prepare fast enough and will have to wait until spring.

Our mule train should make it before the heavy drifts hit the mountain passes. We hope.

No trouble getting across the Big Blue and the Little Blue. Mules and horses swim good.

Ft. Kearny hasn't changed much. Colonel Powell came out and gave us a warm welcome. Yes, he surely has heard about California Gold. Lost 12 troopers just last week who decided they had a hell of a lot better chance getting rich in California than from Army pay.

He looked carefully at the negroes.

"Where'd you get the colored? Slaves or free?"

"They are all free, Colonel, and all work for me."

"Got the papers, Mr. Munroe?"

"No, Colonel. They been free a good while and lost the papers."

"I see. Then my advice is not to tarry long. In fact you might best be hitting the trail while there's still a little light. Good camping just up the Platte at the first cottonwood grove. There's a little spring there and good grazing."

"Much obliged, Colonel. If you ever get to California look us up. You can be assured of a warm welcome."

"I might just do that, Mr. Munroe. My enlistment will soon be up and me and my friends just might head on out there come spring."

About noon the next day we made it to the Pawnee village and kicked up quite a welcome. Hugh Monroe, Aokii'aki and Sopo'aki are famous.

We all got fed, including the Negroes. Then the chief took us three famous folks into the big lodge and we smoked the pipe.

Four men got up and sat in a row across from me. I smacked the noses of the first three with my Spirit Feather and they sneezed and doubled over. Lots of whoops and hollers.

Then everybody quieted down and sat real still, watching the last Indian. I waited a good 3 minutes, and then gave him a good whack. Oh my, Indians surely know how to have fun. They laugh with their whole bodies, just like little children.

After many thanks and a few plugs of tobacco to the chiefs, we pushed on up the Platte toward Fort Laramie, around 350 miles away.

Whenever we could, we camped at a creek or spring with a little pasturage. The Platte is foul tasting and full of silt. In a pinch, throw a hand full of cornmeal in a bucket of Platte water and wait an hour. Most of the silt and bad taste drop to the bottom.

Shot a few deer and antelope. Cooked some, jerked the rest.

Easy ride to Ft. Laramie. It isn't officially a fort yet. The scuttle butt is that the army will buy the fort soon. A few troops are already there but don't seem too happy with the accommodations. They have a little ditty they like to sing:

> *Soupy-soupy-soup, without a single bean!*
> *Porky-porky-pork, without a streak of lean!*
> *Coffee-coffee-coffee-without any cream!*

There was a bunch of poor Indians hanging around the fort, and we gave them a few provisions. Mr. Monroe looked so sad after he finished handing out the corn and beans.

"It's the end of the wild Indian, Mary. Gray Fox once asked me how many white men were coming. I couldn't bring myself to tell him they are like stars in the sky. So glad I saw the wild life before it's gone."

We have three hundred miles to go from Ft. Laramie to the climb over the Great Divide at South Pass.

The second night out of Ft. Laramie we ran into big trouble. While we were asleep, one of the Negroes broke into a keg of whiskey. He quietly followed Harriet out to do her duty, and when she finished attacked her. She is one tough bird, screamed and pulled her knife. We came running.

Jackson was on his back kicking his legs, shrieking and holding his stomach. He looked up at us and pointed to Harriet.

"She knifed me. I didn't do nothin'. She knifed me for nothin'."

"He grabbed me and tried to kiss me. The bastard deserves what he got. **I hope he dies**!"

We could smell the whiskey, so there wasn't much doubt who was lying and who was telling the Truth.

Sammy bent down. "Bad. Real bad. Cut open his stomach."

We carried him back to camp and Sammy dressed the wound best he could. Jackson was dead the next morning. Mr. Monroe called a meeting.

"She didn't have to knife him," one of the negroes said. "Sure, kick him in the nuts, but don't knife him."

"You don't know how scared I was, black man. Especially attacked by one of you."

Oh my, I thought. This is not going well.

Another negro jumped up, "So you knifed him because he was black. You stinkin' bitch!"

The other negroes started to get up, years of rage in their eyes.

Both Sammy and Mr. Monroe got up quicker.

"Stop it!"

"Enough!"

Sammy turned to the blacks. "Ya, you could kill these white folks, right now or in their sleep. And I wouldn't overly blame you. You been taking shit from white folks all your life.

"Except. Except, I been friends with Mr. Monroe a good long while. We fought together, starved together, whored together. I trust him with my life and he trusts me with his.

"Ya, sure, these white women look down on black folk. They don't know no better. They acts up to the lights they got."

One of the blacks stood up, "then get rid of that woman. Send her back to her peckerhead friends." Growls of agreement.

Sammy turned to Mr. Monroe. "How about it?"

Harriet stood up and faced the blacks.

"You are right. I look down on blacks. It was the way I was raised. I'm not proud of that, but there it is. Look, I'll go back if that is your wish, but how about this. From now until we see the Sierras I will cook your food and wash your clothes. You can treat me like you've been treated."

Sammy huddled with the blacks and after some hard words and a lot of loud talk, everything quieted down and he turned to us.

"Agreed."

Kinda dry this time of year so we kept our kegs topped up pretty good. Might find a little water seeping into a hole we dig in a dry creek bed. Maybe luck out and find a little spring marked by a little greenery at the base of a hill.

When we find a place to camp with good grazing, fuel for the fire, and good water we stop for a day or two to let the mules and horses graze and rest. Ya, and we also need a little rest time but it's mostly for the stock. They die quick from overwork and not enough grass and water. Mr. Monroe is a very careful man.

Real dry climbing up the grade to South Pass, so we topped the kegs. It's a nice gentle climb that makes it the best pass over the divide. Well, Mr. Monroe says Marias Pass way up north in Blackfoot country is just as good.

It's about 100 miles to Ft. Bridger from South Pass. Mr. Monroe says it's a beautiful place compared to the dry, treeless country we been through. When we get there we'll rest up the stock and get ready for the final leg of the journey across the hot, dry desert of the Great Basin.

Haven't decided yet whether to take the usual route going north to Ft. Hall and then along the Humboldt to the Carson Valley, or down to Salt Lake and try a new trail cutting straight across a few low mountain ranges.

Bridger got word we were coming and was out in front of his little town to greet us, with most of the town folk right behind him.

~~~~~~~~~~~~~~~~~~~~~~~

I put this in between the pages long after the day we met Mr. Bridger. He wasn't stern like the picture. He cracked jokes and was very friendly and pleasant.

~~~~~~~~~~~~~~~~~~~~~~~

"Monroe, you old bastard. How the hell'd you ever out run all those Sparrowhawks and Sioux?

"Just like Sammy here, you old mother's nightmare, I ran like hell. Good to see you again. Last time was at the Devil's Gate, up the Missouri. Seems like you were running from a bunch of your friends who figured you beat them out of a good bit of peltry."

"Rumors, old friend, nothing but lying rumors. I have always conducted myself with the utmost attention to honesty and truth." Huge uproar of laughter from his neighbors.

"Ah Yes, Mr. Bridger. Your reputation for truthfulness and honesty is well known to all, far and wide." Another big uproar of merriment.

"You're off to the gold I suppose. Well, good luck with that. Seems like everybody around here has the same wild dream of getting rich quick. Got time to tarry? I'd surely like to swap a few tales of long ago when we were young and wild."

"With your kindness and hospitality we hope to spend a couple days grazing and loafing. You ever see Sublet or old Beckwourth?"

"Haven't seen or heard anything about Sublet in a good while. Jim went through here around 9 months ago. That tough, old bastard was carrying dispatches to Monterey. Hasn't been back so he's probably digging gold like everybody else."

"Ya, Jim is the toughest man I ever met. But honest and a good friend."

"You got that right. I miss those old rambling days. How the hell are we standing here talking in the warm sunshine when most all our friends were either left for the buzzards or dropped into a shallow grave?

"Ok, enough of that. There's a fine spot just up the river to camp. When you get settled in, you and the ladies come on back and we'll give you a good feed."

A couple of the blacks got fire in their eyes again, but most just took the expected and didn't get their feathers ruffled.

Mr. Bridger advised us to take the cut-off down to Salt Lake, around the south end of the lake, and straight west to the Carson Valley. There are three narrow mountain ranges running north and south to cross but the passes are low and the water and grazing plentiful.

Not much of a trail yet, but you can see where the passes are from quite a ways off. He wouldn't try a wagon train but we should have no trouble.

"That cutoff is a hell of a lot better than the alkali dust and the devil's heat on the Humboldt. That white sand blazing up from the hot sun boils your brains out. And you'll save at least a couple hundred miles."

We been making good time, even with the stopovers to care for the stock. Been on the trail around a month and a half and covered around 1500 miles, so we been doing around 25 miles a day.

Going to be tight, getting over the Sierras before the snows hit. Mr. Munroe says a good storm can pile 6 feet of snow easy.

Mr. Bridger figures it's around 400 miles or so from Salt Lake to the Carson River, so we should get there around early October. Then over the Sierras to California gold!

Osgar is sticking with us. Not getting bored like with the wagon train. He is so lively, like he already smells the gold. I heard about the pot of gold at the end of the rainbow that Irish elves chase.

I guess he figures his pot of gold is off where the sun sets, on the other side of the Sierras. Wonder what Irish elves do after they find their pot of gold.

Salt Lake is just getting started. The first group of Mormons got here last year and are busy building adobe and clapboard houses, working the fields, and tending stock. Hard workers, cheerful and friendly.

They don't know anything about that route straight across to the Carson River. No time to explore. Best to hurry around the south end of the lake though. Hot, dry and dead.

The blacks let Mr. Munroe barter some pickaxe heads and hammers for a lot of fresh provisions.

The next morning we topped our water kegs and started out. By evening we were around the end of the lake and could see mountains in the distance. Ya, like Mr. Bridger said, it's easy to see the gap.

There are small creeks and springs along the way and we cried with joy when we got to the thick green grass and pine and cedar forests of the first range of mountains. We shot a few antelope and deer along the way. Ate some, jerked the rest

Snow dusted the top of the peaks of the second range and it started to get colder and colder so we broke out winter gear. A foot of snow on the pass of the third range.

And then a few days later, we started across a dry, ugly salt flat. The hot air was full of gray alkali dust, and rags over our nose and mouth helped a little.

What joy when we reached the eastern edge of the Carson Valley. There on the other side of the wide canyon were the glorious mountains we had come so far to see. The High Sierra Nevada Mountains of California.

Thank God we didn't do the Humboldt.

Harriet's servitude is over. She fulfilled her duties with vigor and a good will. There were no complaints from the Negroes and they treated her much better than they've been treated.

She was grateful for that, and even more grateful to finally see the Sierras. Her hands are getting a mite sore from all that cooking and washing.

We camped on the Carson and the next day headed south, upstream. Took a fork to the north west, climbed to the headwaters, and over the pass into California. Not so tough.

A few miles down the creek we saw a big lake at the bottom. Never heard of any big lakes in gold country, so we must still be on the east side of the Sierras. It's getting colder and colder, and a wild winter storm could happen any time.

Half way down we saw smoke from campfires on the edge of the lake. Must be white folks. No Indian would ever make a fire that smoked like that.

We hollered hellos at the first camp and got hellos hollered right back at us. They are hunters and fishermen, gathering provisions for the mines. If we stick around a couple days, we can follow them when they go back to Dry Diggings.

Snow not deep enough to stop us yet. If it looks like a big storm brewing we'll head right on out.

"How far to the gold fields?"

"It's about fifty miles to Dry Diggings, the heart of the gold country."

We hunted and fished right along with them, and a couple of days later followed them over the pass, into El Dorado.

7
DRY DIGGINGS

On the other side of the pass we dropped down into the headwaters of what our friends told us is the South Fork of the American River. We saw a few prospectors panning the gravel and a little farther down there were three men and a dozen Indians working two holes in a gravel bar.

We stopped for a look and one of them explained what they were doing.

"Most of the gold is maybe 6 feet down. The last foot before bedrock is the best, and we run it through the cradle over there. Then when we hit bedrock and there are cracks in the shale, we use our knives to pry out the dust and nuggets. Don't know about this hole yet. The last one only panned out fifty bucks."

"How much for a good hole?" Mr. Monroe asked.

"Well, old Hunter Beck and his crew swore they got over $3,000 out of a hole once. But that's either real good luck and real good lying. Good holes run around $200. That's a hell of a lot of money considering back in the states I was only making around $400. a year."

Oh my, how our mouths did water and our eyes did sparkle.

We just about stopped right then and there, and started digging. Osgar surely wanted us to dig and kept pointing to a spot a couple hundred feet down river. But we decided best to go on to Dry Diggings and learn more about the lay of the land. We can always come back.

I went over and signed one of the Indians. ~~you work whites?~~

~~yes~~pay little~~we hunger~~

~~many Indians work whites?~~

~~yes~~children hunger~~whites kill deer~~whites kill fish~~whites kill Indians~~

We passed two small Indian villages, with six or seven brush huts. Mostly sad and half-starved women and children.

———————

Dry Diggings is the ugliest, messiest town I ever saw. And it stinks. Oh good Lord, how it stinks. Animal droppings everywhere, and ya a few human droppings as well. Trash tossed everywhere. Scum covered pools of water in the middle of muddy streets. I even saw a couple big rats run into a tent.

~~~~~~~~~~~~~~~~~~~~~~

Well I got this picture of Hang Town a few years later and stuck it in here. It shows Hang Town around 1850 after the hanging changed the name. Couldn't find any earlier picture.

Looks a whole lot bigger and cleaner than when we arrived in late 1848. Mostly tents when we arrived. In this picture it's log cabins and clapboard shacks. You can see all of the stumps on the hillside, the trees cut to build the town. If you look real close you can see the hanging tree, the big oak just up and to the left.

~~~~~~~~~~~~~~~~~~~~~~~

The only clean looking place is a new clapboard shack on the edge of town with two smiling women sitting beside pretty flowers, waving and smiling at the gentlemen as they pass. A black woman puttered about, keeping everything clean.

Betsy and I wanted to go on over and say hello, but decided best to wait and learn a little more about the town.

There are a few ladies around who are not smiling and waving, just washing and cooking. They aren't having much luck keeping up with the dirt and mess, but I bet the fellows are happy with the cooking.

Ya, the town's a mess, dirty and smelly, but full of high spirits and excitement and a lot of log cabins and clapboard shacks are going up, leaving a bunch of stumps on the hillsides.

A few Indians are walking around and helping build the cabins and shacks. Their brush huts are just north of town.

The Negroes are camped in tents to the south of town. Sammy, his friends and their 25 mules went over to say hello and learn the lay of the land. He once told me the first thing a Negro does when he hits a strange town is find a brother and learn where he can walk, talk, eat and fornicate.

Osgar is running and jumping up and down, pointing here and there. He gave big kisses to the ladies as he went by.

Our band of 50 mules, Mr. Monroe, an old squaw and seven mighty pretty ladies started down main street and attracted a good bit of attention.

Three or four men shouted out greetings and Mr. Monroe waved and shouted back. Then a big rough looking fellow came running over, grabbed Mr. Monroe off his saddle, and gave him one hell of a hug. He is the toughest and scariest man I ever saw and has at least 3 scars running across his face.

~~~~~~~~~~~~~~~~~~~~~~~

I got a picture of old Jim years later and put it here between the pages.

~~~~~~~~~~~~~~~~~~~~~~~

"Monroe! Monroe! Haven't seen you in a coons age. How the hell you been?"

"Well I'll be damned, Jim Beckwourth, you old scallywag. Mighty good to see you again. How you been?"

"Mighty fine, thanks. You got some mighty fine women there. Always knew you were a ladies man. Didn't know you kept a harem."

Then he suddenly stopped and his eyes got wide and wild.

"Aokii'aki!", and went over and they touched foreheads.

"You have seen much and still live. Your medicine is strong."

"No, Aokii'aki, your medicine is strong. Around my neck is the medicine bag you gave me many years ago. I have seen you in dreams and in the waters and heard you in the wind watching over me."

"Yes, it is so."

"You have a place to stay?'" turning to Mr. Monroe.

"No, we just hit town."

"Then come on out to my place, and meet my wife. I have good water and pasturage. Plenty of room for all of you, your stock and your Negroes. I saw Sammy go by. Always had a lot of respect for Sammy. Smart, strong, loyal, and one hell of a tracker and fighter."

"Mighty obliged, Jim. Lead on." The negroes and their mules rejoined the group on our way out of town.

"Good to see you again, Sammy."

"Same here, Mr. Beckwourth. We had some wild times up the Wind River."

"Yes indeed we did. You made off with a whole lot more peltry than I did. You are one hell of a peltry packer."

"Learned from Mr. Monroe."

Mr. Beckwourth's place is five miles out of town in a beautiful valley with a good spring and pasturage. We took off the packs and put them in the barn.

Oh my, how the mules did kick up their heels as they headed for the high grass.

He invited everybody into his large log cabin, including the negroes. Not many people know that Jim's mother was a mulatto, and his father was Sir. Jennings Beckwerth, an English nobleman.

His wife saw us coming down the trail and soon had a big meal hot and ready.

"May I introduce my wife, Maria."

She's Mexican and pretty. Kinda young for old Jim but doubtless a warm comfort on a cold winter night. We exchanged pleasantries and sat down to the best meal we've had in a good long while.

The old timers spent the rest of the evening reliving the wild life of long ago. If even half of Jim's tales are true, he must be the greatest Indian fighter of all time. He even said he was once the head chief of the Sparrowhawks, the tribe the whites call the Crow. Mr. Monroe backed up his stories but he was maybe just being a good friend.

After many a fine tale, and more than a few drinks of wine and brandy they talked themselves out and we all went to bed. Sammy and the negroes sleep in the barn and I was lucky to get a nice place on the soft rug in front of the warm fireplace.

The next morning over breakfast we talked about Dry Diggings and how to get rich. Mr. Beckwourth said that only a few miners were getting rich.

"Most are having a hard go of it. The cost of food and mining equipment in gold dust is out of sight and what the miners don't eat or use, they blow at the gaming table, saloon, or friendly women. And there surely is a whole lot of room for friendly ladies," with a quick look at our ladies.

"Gold is easy to get and hard to keep. The only sure way to get rich is to sell goods and services to the miners."

We're free to stay at Mr. Beckwourth's until we get settled, but we decided to leave most of the mules in his pasture and our merchandise and blacksmith equipment in his barn and go back to Dry Diggings and scout the territory.

Sammy's two blacksmiths and their helpers know they can make a pile of money working the iron, but they're first going to try their luck getting rich quick.

The negroes in town have claims on a small creek not too far from town and Sammy and his buddies are going to join them. Mr. Monroe and I will sell the iron goods they brought from Independence in our future emporium for 20% of the price we get.

Betsy, Harriet and Molly don't much care to cook and wash. They're going to talk to the two ladies who got here first, look over the land, and figure how best to set up their house of fun.

The three nuns don't know what to do.

Good thing we have tents because there was one hell of a downpour that first night in Dry Diggings. And the rain didn't stop. It just kept on all night, and all the next day. The sun finally broke through early the next morning.

Good Lord, I thought the town was bad when we first arrived. Just try coming to town after a good rain. Mud everywhere, and all that awful smelly garbage that was dropped and thrown away just gets picked up by a river of mud flowing through town.

And people get all that filthy dirt on their boots and trudge them into their tents. The only wooden sidewalks in town are in front of the saloon and the two smiling ladies.

Mr. Beckwourth warned this is the rainy season. The heavy clouds come storming in from the Pacific, hit the Sierras, can't get over the high mountains, and dump. And oh my God how they dump. For sure the passes are all snowed in for the winter.

The dry season starts in late Spring. Pleasanter weather but then the miners start running out of water to rock the cradles. That's why the place is called Dry Diggings. They either quit until it rains, or cart paydirt to water.

Aokii'aki and I went to visit the Indian encampment north of town. There were 12 or so brush huts, in good condition with squaws and children sitting in front. No braves except for an old Indian sitting with his squaw in front of a brush hut.

I gave the sign of greeting.

The old Indian smiled and returned the greeting.

~~Spirit Woman welcome~~sit~~talk~~

~~thank you~~I Coyote Woman~~

~~yes Coyote Woman~~remember~~in dream~~me White Fox~~

I brushed my spirit feather across his forehead.

~~yes~~dream~~ touch forehead~~

~~are your people well?~~

~~no~~we work get gold~~whites steal gold~~whites hunt kill men~~end of Indians~~

~~what you need?~~

~~work~~food~~children hunger~~

~~women want work?~~

~~yes~~men die~~children hunger~~

"Aokii'aki, let's hire a woman to help cook, wash, and clean."

"We hire two. Children hunger."

The old man gathered the women who wanted work and Aokii'aki picked two.

Later that day Mr. Munroe and I visited the three merchants in town to learn prices and what's selling. Mr. Monroe knows one of the merchants, Gus Wheatley, so we got straight up information. The other two seemed friendly enough even when they learned we were setting up, so guess there is enough business for all.

Mr. Wheatley said ok to put up our mercantile tent on his vacant lot right next door. Guess he figures he'll get some of the miner's business when they come to our opening.

The three nuns don't want to wash, cook or whore, so they're going to dig gold. Mr. Monroe went back to Mr. Beckwourth's and packed three mules with a tent, provisions and gold pan, picks and shovels for the ladies. He'll show them how to carve the handles. The nuns bought a cradle while we were gone.

I've mentioned a cradle a few times, so maybe best describe one. It's also called a rocker.

~~~~~~~~~~~~~~~~~~~~

Well I tried to use words but never could get it right so got a picture a few years later. I've stuck it in here and added a pretty good description how it works.

You throw a bucket of pay dirt into the hopper on top. It's a box with a mesh bottom that collects the big stuff and lets the rest fall through. You throw buckets of water into the hopper and rock the cradle with the handle.

The pay dirt falls through the mesh and the gold and black sand get caught behind the riffles. Those are two 1-inch high pieces of wood spaced six inches apart or so, nailed across the flow of the water running down the bottom of the cradle.

Now and then you dump the rocks from the hopper and pan what you caught behind the riffles.

Cradles work the dirt so much faster than a gold pan. One man can operate a cradle but three are best. One shovels and feeds the hopper. One throws the water. One rocks the cradle.

~~~~~~~~~~~~~~~~~~~~

Osgar told me to go with them so he could tell them where to dig. So off we went, three nuns, an innocent little girl, an Irish Elf, and three pack mules.

Osgar found a nice place up a ravine and the nuns set up their tent. If they find much gold I told them to keep their mouths shut. Otherwise a whole mess of miners will trample them into the sand.

On the way back to town I went to say hello to Sammy and the Negroes. They were hard at work digging their 10 X 10. Osgar pointed to a spot up creek and I took Sammy aside and told him about Osgar. Sammy already knew about Osgar, but didn't know about his nose for gold.

We walked up to the spot and he marked it with a couple of stones side by side. They'll go there when they finish the hole they started. He knows not to tell anybody if they strike it rich.

Mr. Monroe painted our mercantile tent a bright yellow to attract attention and help keep water from leaking through, and painted a pretty sign over the entrance in big white letters, PLACER EMPORIUM.

Most all the miners come to town Sundays to spend their gold on provisions, gambling, drinking, and fornicating. There are no church bells to dampen the revelry. So we carefully arranged the merchandise and early on a sunny Sunday in early December, 1848, we opened for business.

We gathered quite an eager crowd outside, and did they ever rush in when we opened those flaps.

What a day. We sold nearly $4,000. worth of our merchandise and about $1,500 of the negroes' iron goods We had to push people out and close the flaps when it started to get dark.

We have oil lamps and candles but we're tired and they will come back tomorrow. We have plenty of merchandise left.

Mr. Munroe and I have to work hard and fast if we want to exchange our round, yellow tent for a proper mercantile establishment before the hoards come up the creek and over the pass next year. We hear that newspapers back in the states are full of grand stories of golden glory. A few early birds will be coming around the Horn and across the Isthmus.

Most are going to be disappointed. All the easy diggings will be mostly gone, and there will be nothing but hard, dirty work for the gold. Then after paying for supplies, and a little entertainment at a saloon, gaming table and a visit or two to the house of the soiled doves, most will have nothing to show but a sore back and sour dreams.

Oh but years later it will all be worth it. They can shout out to the whole world, "I was a forty-niner!"

Betsy, Harriet and Molly went to see the two smiling ladies. They do not care to share and there was a minor brawl. The smiling ladies only have a cottage. Betsy and I want to have at least a low end house, but don't have enough money to buy land and start building.

It took a big chunk to fund the high end house in Independence, and I put up half the funds for the goods and trip to California. I can provide some money from my half of what we make at the emporium but have to save most for a buying trip to San Francisco.

The three girls don't think much of selling their sweet merchandise in tiny, dirty tents, so decided to go see how the nuns are doing. I told Betsy about Osgar and his wonderful nose, so she wants him and me to go with them.

I suggested she and the girls wait a couple of day until selling quiets down at the Emporium and then we can go together. They'll wait and help with the store in exchange for the outrageous salary of twenty dollars a day. Back in the states that's what you get for a whole month's work.

The next morning after the opening we were graced with a visit from the leaders of the community. Mr. Wheatley was among them. We were welcomed by Mr. Jonathan Perkins, leader of the leaders of the community.

"Mr. Monroe, we are pleased to have you join our community and hope you will grow and prosper. We would be grateful if you would agree to join our Truth and Justice Committee. As you know, there is no formal law enforcement here. We miners and merchants must provide our own.

"As yet there have been only a trifling amount of thievery and claim jumping and we wish to keep it that way. We meet once a week in Mr. Wheatley's store to go over the affairs of the community. Will you join our company?"

"Gladly. I have noticed a gratifying level of honesty in the camp. People feel free to leave tools and provisions lying about undisturbed. But as you know, next year as people pile in over the mountains that will surely change."

"We greatly fear you are correct. We hope knowledge of our presence will diminish the magnitude of the change."

The next morning I was graced with a visit from Mrs. Wheatley who invited me to meet a select group of lades of the community for tea and tarts. I have never been invited to a meeting of the select ladies of the community for teas and tarts before.

I was very surprised and pleased to find that they are actually nice people, not the stuck up, prissy prudes I always imagined. Well yes, there was one Mrs. Grundy in the group, Penelope Perkins from Boston, wife of the leader of the community. She wished to know my background and I let her have it.

"My dear father is a prominent banker in Chicago and my mother is from a good Baltimore family. They ensured my education but I fear are not at all pleased that I decided to leave home for the rough life. I am so glad Mrs. Perkins that I have found such fine people as yourself and the other ladies who have so graciously invited me to share tea and tarts."

Oh, I was so tempted to shout out the truth.

I'm a whorehouse bastard!

Just to see Mrs. Penelope's face change into a horrified prune. But I stopped myself. Mr. Monroe and I wish to maintain a respectable establishment in this community.

I apparently passed. I am now a member of the Ladies Guild of the Golden West. Thank you Polly for all those books you let me read. I graciously accepted membership in the committee planning the opening of a school. They also have a committee for the needy, and a church fund.

Yes, the ladies are a bit stuffy, I doubt much foul language has ever passed their ruby lips, but are genuinely warm and friendly. Maybe they loosen up when they get to know you better. Hmm. On the other hand, maybe it would not be such a good idea to loosen up and get to know each other better.

It wasn't far to the nuns' claim. We saw their two blue tents long before we saw them. They must have painted them to help keep out the rain. They are surely working hard. Mary Ellen feeds the hopper. Sophia throws the water. Bethany rocks the cradle. They all shouted hellos when they saw us.

Mary Ellen ran into their tent and came out with a smile and a pouch. They have only gone down maybe 4 feet and already have 10 ounces of gold. They are so happy and proud. They have no idea what they are going to do with all that money, but for sure a whole new wonderful life is just around the corner.

How about Grand Ladies promenading down the avenues of San Francisco.

Osgar kindly showed me a spot for Betsy, Harriet and Molly up the gulch not far from the nuns. I'll help dig until I have to go back for the buying trip to San Francisco.

Harriet and Molly have agreed to invest in our fun house. They will give Betsy most of their share of the gold we dig and once our establishment gets going Betsy will pay back double their investment.

Harriet will tend the bar and only entertain guests of her choosing. Molly will get the best room in the house to entertain her guests.

I told them about my membership in the Ladies Guild of the Golden West. We decided I would take a very low profile in the workings of our somewhat less refined lady's group. I will let them know the sentiments of the straight world and do what I can to lessen any negative feelings toward our future house of fun.

It's hard, dirty work down in the hole shoveling dirt into a couple buckets, especially when there's water from the rain. Betsy dumps a bucket of pay dirt into the hopper while I'm filling the other one. Harriet throws the water from the creek and Molly rocks the cradle.

We noticed that the nuns frequently changed tasks and soon found out why. Different jobs use different muscles.

The rains made the work very unpleasant, but now and then the sun came out and the gold flakes made such a wondrous display, shining in the sunlight.

After only a couple days Osgar's splendid nose proved its value once again. Eight ounces and we're only down 3 feet.

A few nights after we arrived I heard a gunshot and screams. Betsy and I grabbed our pistols and pushed open the tent flap. Two of the nuns were trying to crawl out of their tent but a couple of men pulled them back in.

We ran down and into the tent. Mary Ellen lay off to the side face down and the two men were trying to rape Sophia and Bethany.

"Get off them you bastards!"

They looked up and laughed.

"Hey Bart, we got two more after we finish these."

Must be too dark to see our guns.

"You got two seconds to get the hell out of here!"

They laughed again and one of them got up and started toward us. We shot him through the heart. The second man jumped up and ran out of the tent. The coward got it in the back.

Mary Ellen is dead. We lay next to the nuns the rest of the night.

Next morning we put the bodies of Bart and his buddy on a couple of mules, went quite a ways from camp and dumped them for the buzzards and coyotes.

We found their two horses near the camp. The saddlebags had over 17 thousand dollars and around 26 pounds of nuggets and dust, along with a load of beef jerky, chewing tobacco and etchings of naked women. Two rifles and a gun belt were slung across the saddle horns.

We heard desperados were roaming the hills robbing miners. Well, they been real lucky getting all that money and gold, but their luck just ran out.

We took off the saddles, hid them and the saddlebags in the bushes, and gave the horses a good whack. They ran off over the ridge.

When we got back to camp we carefully packed the etchings. They will grace the walls of our parlor and will surely increase the speed that the dears choose their lady friends.

We wrapped Mary Ellen in a soft red blanket and buried her in the middle of a beautiful cedar grove. Sophia and Bethany bowed their head in prayer, and then we all just stood for a few minutes, shedding a few tears and saying our final goodbyes.

Take care, Sister Mary Ellen, old friend.
Fare thee well.

We're not going to tell anybody what happened. If anybody asks, Mary Ellen got hit by a big rattler.

Sophia and Bethany are joining our enterprise. The two nuns figure digging dirt may get a little old after a while so they want to open up new opportunities for investment and employment. They took the rifles and pistols and are going to learn to shoot.

We decided it's faster for everyone to work one hole at a time. We alternate filling the buckets, dumping the dirt and water into the hopper and rocking the cradle.

We reached the shale bedrock in a week. Beautiful, shiny nuggets in the cracks. Altogether we gathered 4 pounds out of that hole. Betsy and I figure that and the 26 pound of gold and cash from the two bastards are more than enough to buy the land and start building our establishment for the Ladies of Joy.

So, early one morning, Betsy and I headed for town. We decided to call our fun house THE FUN HOUSE.

We looked all morning and finally found a nice spot for sale on a little hill, down the gulch a little ways from town. I let Betsy go talk to the guy who owns the lot. Later she told me that he drove a hard bargain and refused her final offer, but then called her back when she walked away.

While I was working the dirt, Mr. Munroe looked around for a site for the Emporium and settled on a nice location downstream on the edge of town not far from THE FUN HOUSE. There is a bit of high ground there, away from the floods and that's the direction the town is going to grow.

An old guy living in a tent owned the land. The rains were not helping his lungs any, so he sold for a good price and went back to Sacramento City. Not sure that'll be any better for his lungs, but whatever.

With all the new people soon flooding in, I'll bet next year our PLACER EMPORIUM will be in the center of town.

Mr. Munroe is going to start with a large room in front for the merchandise and a smaller room in the back for eating, sleeping and storage. We'll build strong so we can add a second story and wings later on. Mr. Munroe has built quite a few clapboard buildings in his day and knows how to build straight and true.

Betsy and I are going to build a nice parlor with a bar and hallways with small rooms going off left and right. Like Mr. Munroe, we're going to build strong to later add a second story. We'll have a simple bar and simple furnishings at first, and get more glorious as we prosper.

Both the Emporium and the Fun House will have a stone fireplace and a cistern to catch rain water. The cistern for the fun house will be extra large and we'll hire Indians to haul water to keep it full during the dry season. We want clean girls.

Mr. Monroe kindly agreed to oversee the construction of both the House of Fun and the Emporium.

If anyone wonders at the apparent friendship with soiled doves, Mr. Monroe and I will tell them we shared much danger and hard times on the trip across the mountains and deserts and owe our lives to Betsy and Harriet.

The two Indian women Aokii'aki chose to help with cooking, washing and cleaning are cheerful and hardworking. I talked to Betsy and she'll hire Indian women to keep the Fun House clean and the clothes and linen washed.

Look at me. I'm quite the business woman. Mr. Monroe and I are 50-50 on the mercantile establishment and I'm a 50-50 silent partner in the future best whorehouse in town.

Not good weather to build but a lot of labor available. Many miners run out of luck and grab any job they can get to keep hope alive until they can earn or beg a new stake. And even many lucky miners can't work their claims when rain and high water swamp their diggings.

Mr. Munroe has to continue the fiction that Sammy and the other five negroes belong to him. A Negro cannot testify against a white man, called the Right of Testimony, so free Negroes can be attacked and claims and gold stolen with no way to fight back.

Sammy and friends entrust Mr. Munroe to hold most of the gold they mine and what they sell at the Emporium, and to help ensure that nobody jumps their claim. Well the Negroes are tough and armed, so only a fool would try.

Using the Negroes' gold Mr. Munroe and Sammy became owners of a lot for the smithy, right on the boundary between the blacks and the main town. A Negro's right to own land is also in dispute.

Betsy and I went back out to our claim to help keep the gold flowing for the building and to add to our fund for the buying trip to San Francisco to buy essentials for our glorious Fun House.

I shall not dwell upon the cholera that came to town. It is a horrible disease, strikes quickly, and kills quickly. Seems to happen most to very unclean miners who eat poorly.

It also seems to happen most often to people who live at the lower end of town, the part of town most subject to the mud flow of human and animal waste. We have built on high ground and are not subject to that dirty wash of water.

The Ladies Guild helps the sick and the dying. The ladies are not afraid to close a dead man's eyes, or too prissy to shake filth from the bed clothes. Yes, even Penelope Perkins. She might be Mrs. Grundy, but she's tough and doesn't mind getting her hands dirty to help a body in need.

8
THE SOILED DOVES OF SAN FRANCISCO

After the roof, sides and wooden floor were completed for the Emporium and Fun House, Mr. Munroe and I started off early one sunny day in late December on a buying trip to San Francisco.

Mr. Munroe is buying for our mercantile exchange, and I'm buying for the Fun House. I have a good bit of gold from the desperados and gold we dug. The Fun House needs wine, whisky, glassware, linen, cleaning supplies, cutlery, bedding, and lots of red calico for draperies and curtains. I'm afraid Harriet's French Silk sheets will have to wait.

The Negroes are sending gold with Mr. Munroe to buy iron bars and two bellows, or enough cloth and leather to make them. They have hired Indians to help work their claim and are starting to build their smithy.

Mr. Wheatley came with us to replenish his own stock and show the best places to buy.

It's an easy sixty miles to Sacramento City. The Negroes will meet us at the dock with the mules when the El Dorado returns from San Francisco in five days. Eight of the mules are Mr. Wheatley's. Twelve for the Emporium. Eleven for the Fun House. Five for the Negroes.

We left our horses at the stables and walked around Sacramento City. It's a rough town. Mostly tents cover the river banks and hillsides, but a lot of cabins and clapboards are starting to go up.

A big clatter of Whites, Mexicans, Chileans, Blacks and Indians hammering and sawing up a storm. A loud riot of squeaky wagons and carts, and a whole lot of drinking, gambling, singing, yelling and cursing.

We noticed a couple of tough guys watching us and hurried back to the dock. When the steamer blew the whistle we got on board. It's 100 miles down River to San Francisco.

We docked at Central Wharf where the steamship company has their warehouse. The goods we buy will be delivered and stored there.

Mr. Wheatley took us to a boarding house just off the plaza, close to the commercial district where we'll do most of our buying. Had a great dinner and then sank into a wonderful soft bed with cleans sheets and warm wool blanket. Heaven.

Prices are high but we're good bargainers. Mr. Monroe will easily get twice what he's paying. I bought the essentials for the Fun House and a few nicknacks and fancy cloth to make the house pretty.

We spent three days buying and then had a whole day to just wander the streets of San Francisco before the El Dorado heads back to Sacramento City.

So much energy, so many people, so many carriages and wagons, all rushing about like ants on a kicked hill. Smell not quite as bad as Dry Diggings but pretty close. There are outhouses but a lot more animal droppings, and most of the population appears to have a bitter hatred of soap and water.

Talked a while to Blue Bell, madam of a low end house. No parlor house yet in town, and only one high end but she expects that will change real soon, what with all the new people crowding into town, and all that gold starting to flow back from the Mines. She has plans for a high end house, and who knows, some day she might become the madam of the first parlor house in San Francisco.

Got lost for a while in a bad part of town. A few cottages, cribs and street walkers. A fellow mistook me for a whore and tried to push me into an alley. Bad mistake.

Then I wandered into China Town. Beautiful fabrics, linen, crockery, glassware and wine. Still had gold left so bought a load for the Fun House at very good prices. The Chinese have just arrived in San Francisco and are eager to sell their wares for gold dust.

I noticed a young, very pretty girl watching as I went from place to place bargaining for the Chinese merchandise. She was tall and slender with an abundance of all of the feminine attributes that make intelligent, respectable men go crazy stupid.

I do believe the good Lord above neglected to provide enough blood for the poor dears to work their brains and manhood at the same time.

She came over just as I was about to leave China Town, bowed her head and handed me a piece of paper.

"Help. Please."

~~~~~~~~~~~~~~~~~~~~~~~

Ah Choi. Her name is **not** A Toy

~~~~~~~~~~~~~~~~~~~~~~~

It was an order for her to appear before Judge Baker that afternoon. A letter from her husband in Hong Cong demanded she be returned to him.

"No marry. No return."

We looked into each other eyes and saw intelligence, toughness, and spirit. A bond was formed, and I saw a grand lady in finery strolling down the avenue in a time soon to come.

I went with her to court and told the judge she swore that she was not married to this man and wanted to stay here in this wonderful land of opportunity. He seemed hesitant so I added that with his permission she will return with me tomorrow morning to help in my mercantile exchange in Dry Diggings.

After court she took off the necklace she was wearing, bowed her head again and gave it to me. It's jade with a golden dragon.

"Please. Go Dry Pigs," and I agreed to take her.

We got up real early the next morning and watched our goods loaded onto the El Dorado. Arrived Sacramento City mid-afternoon and met the Negroes and mules at the dock. It was dark by the time the mules were packed but we started up the trail anyway and arrived well into the night, tired and very happy to be back.

―――――――――――

We brought back a big parcel of lemons for the scurvy outbreak. The Hudson's Bay Company always kept them handy for their employees. The Royal Navy learned long ago that lemons stopped scurvy among the Sailors of the Line. The Indians have no problem with scurvy because their diet has a lot of fresh or dried fruit.

We Ladies of the Golden West spent the day going around Dry Diggings giving out the lemons to anyone with bleeding gums, compliments of the Ladies Guild and the Placer Emporium.

A few didn't believe lemons were the cure but most took a lemon anyway. Tiredness, aches and pains, especially in the legs, along with the bleeding gums, were getting worse and worse. They were grateful for any help they could get.

―――――――――――

Betsy and I plan to open the doors of THE FUN HOUSE, along with open whiskey kegs, open wine bottles and assorted other open items on the first day of March, 1849.

Bethany and Sophia, the two nuns, have had enough mucking in the mud and don't want to whore, so they took their share of the gold, quite a tidy sum, and headed for San Francisco.

They don't know what they will find, but are surely looking forward to the next chapter of their full and adventurous lives.

Jennie and Rosie, the smiling women in front of their clapboard cottage, decided best to come join. They can't compete with a lot of pretty woman and a parlor with a bar, and they were getting a little lonely for other members of the sisterhood.

We were glad to get them because they know a lot about the town such as the prices to charge and who are the good fellows and who are the bastards. There was that little brawl when we first arrived but that just showed they have a little spunk.

Betsy will run the day to day operation and we'll make important decisions together.

Big problem. We are a little shy of working girls. Just Harriet, Molly, Rosie and Jennie. We figure we need around 15 to start. Jennie told us she has friends in Sacramento and San Francisco who will jump at adventuring for their pot of gold.

Betsy went over the four requirements of a good whore that Polly gave us. They must enjoy sex, have a pleasant and warm personality, not the first time selling, and keep themselves clean. Jennie agreed. She and Rosie love flowers and are going to plant them in front of the Fun House, just like they had in front of their clapboard shack.

Betsy also went over how to protect the girls from disease. A guest's privates must be washed by his girl before engagement. If a girl sees sores, lesions, or a rash, she must immediately report to Betsy. And, she must quietly report any on herself or another girl.

Jenny raised her eyebrows, "No, our guests do not complain. They are afraid of disease just as much as we are. Our reputation gets around, and guys with signs of disease don't come. And the dears so love getting lovingly washed, and our girls do not disappoint."

So tomorrow, Betsy and Jennie are off to find young adventurous women. Harriet will keep everybody working hard getting ready for our grand opening.

Except they didn't leave. We had our first murder. A merchant on the edge of town was robbed and killed. The shot woke the town, and the Truth and Justice Committee quickly organized the men.

Sammy's many years as a tracker were more important than the color of his skin, so he led the chase.

The two murderers were caught in a box canyon and brought back to Dry Diggings. A trial was held late afternoon beneath the big oak in the center of town. The Committee heard everybody who wanted to speak.

Two boys identified the two men as the men who ran away and Sammy said their tracks led from where the merchant was murdered.

The vote of the citizens of Dry Diggings was unanimous. Ropes were slung over a thick branch of the big oak, the murderers hands were tied behind their backs and handkerchiefs tied tight across their eyes. They were put on their horses, nooses tightened, and a sharp whack sent them to their reward.

It was a horrible sight because they didn't die right away. They gagged and kicked for quite a while. Should have shot em in the head.

Betsy and Jennie left the next morning. Molly will meet them at the dock in Sacramento City in 5 days with horses and mules to pick up the soiled doves and their belongings.

They kept the corpses hanging from the tree until the smell got too bad. Everyone hopes the display of what happens to thieves and murderers in our town will knock a little sense into people tempted to get their gold the easy way.

Well it may or not knock sense into anybody's head, but it surely changed the name of the town from Dry Diggings to Hangtown.

The easy going town changed after that. More and more shady characters were arriving and people began to be a whole lot more careful what they left laying around.

Mr. Munroe quietly dug a hidey hole beneath the floorboards in our back room for most of our gold, and put a bunch of boxes on top. He put some gold in a small box in the store next to the brass gold scale.

If a thief comes with a gun, Mr. Munroe will put on a big show of fear for his life, and terrible anguish at the loss of all his gold as he points to the small box on the counter.

The thief had better be smart and quick or he won't make it back out the door.

Mr. Barkhurst just started building the first hotel in Hangtown. Our little town is fast becoming a big town.

~~~~~~~~~~~~~~~~~~~~~~

I found an old picture of Mr. Barkhurst's clapboard Placer Hotel a few years later and stuck it in. You can see the big oak behind the hotel.

~~~~~~~~~~~~~~~~~~~~~~

Betsy, Jennie and 14 pretty ladies came riding into town. They had no trouble getting the girls and could pick the best because so many wanted adventure and gold.

The Fun House has corridors going off each side of the parlor for a total of 20 small rooms. While they were gone Harriet worked hard getting them ready.

Two local carpenters are building the beds, tables and chairs. They work cheap because of the benefits. Yes, you probably remember Polly saying that dangling pretty women turn men into little, hungry boys reaching for the cookie jar.

Ah Choi helped with the mercantile for a while but then begged to work at the Fun House. She is a real fast learner and Betsy is teaching her letters, sums and how to keep books. Ah Choi listens and watches very carefully all the workings of getting our fun house ready.

After the ladies settled in, they walked around town wearing their finest. It was two weeks before our grand opening and we kindly let the carpenters, and select members of the community preview the merchandise.

Carpenters free. Members of the community pay extra for the privilege, except for members of the Committee for Truth and Justice. They get special treatment. Mr. Perkins is easily satisfied.

Penelope Perkins either doesn't know how much fun her husband is having or is just primly looking the other way.

Ah Choi is wonderfully milking the men with her graceful shyness. She plays the role beautifully.

When Sammy's smithy was close to finished there arose the question of what to call the black section of town. Some wanted to call it Nigger Town.

Things got pretty tense. Some of the blacks, mostly Sammy's friends, were shouting back "Peckerhead Town" when they heard "Nigger Town" thrown at them. Just yesterday both sides drew their pistols and waved them as they shouted at each other.

The Committee held a town meeting to decide the matter. The strong speeches from Mr. Beckwourth, Mr. Munroe, and many citizens of the town, along with the memory of Sammy's competent service to the Committee tracking the murderers, easily carried the day.

Hammer Town was chosen because of the smithy. Good advertisement for Sammy's friends. There are only two other smiths in town, and Sammy says they aren't very good.

9
THE FUN HOUSE

The grand opening of THE FUN HOUSE was a grand success. Sold a lot of booze at the bar and a lot of women in the rooms. We hired a fiddler, and the building shook from the stomping feet.

You won't believe it but we sold over $2,000. worth of booze and girls opening night.

Ah Choi is our most popular and made a pile.

We hired two more Indian women a couple of weeks ago to tend the kitchen. It's in a small, separate building just a short covered walkway away. It isn't part of the main building to keep down the danger of fire. They cook for the girls and the hired Indians, and make trays of tasty bites for the customers, well salted to keep the drinks flowing.

Had a little fight start between two men over Molly but our two carpenter bouncers took care of that. We got to know Mike and Johnny petty good.

They are husky, mean looking men but gentle with the girls. Three of the ladies go out of their way to keep them happy. I help with Johnny now and then.

Everything is good. The nuns are probably Grand Ladies of San Francisco by now. Mr. Munroe and I are getting richer and richer and soon will have to make another buying trip to San Francisco. Sammy and friends are gathering the gold from their diggings and smithy.

And Betsy and I have just opened the first house of fun in Hangtown.

Well yes, I made that all warm and rosy. It helps me forget Dancing Deer and Sister Mary Ellen, and stops me from looking too far into the future, wondering how many more "Fare thee wells!" there will be as I travel this strange road called life.

I keep remembering that beautiful goodbye Aokii'aki 's father gave her and the tribe just before he died.

A little while and I will be gone from among you,
whither I cannot tell.
From nowhere we come, into nowhere we go.

What is life?
It is a flash of firefly in the night.
It is a breath of a buffalo in the wintertime.
It is as the little shadow that runs across the grass
and loses itself in the sunset.

########################

And that is all I have been able to recover. I looked and looked online and around town but have not been able to find any way to recover the rest of the story. Let your imagination flow freely. What happens to Mary? What happens to Ah Choi? And does Bottomless Betsy get that grand Parlor House with a winding staircase and French Silk sheets overlooking the Golden Gate?

########################

####################

A couple of months later, after a long search, I found a magic light that was able to recover a little more of Mary's story.

####################

The next morning the four Indian women we hired in addition to the two cooks had to work mighty hard cleaning up the mess and washing the bed sheets and linen.

Betsy Ann summed up the money and gave each girl half of what they earned. The girls usually get paid Mondays at lunch, but this was a special time. The girls made more last night, including tips, than they ever made in a month in Sacramental City or San Francisco. And they get free room and board.

A little while later Ah Choi went back to San Francisco. She has enough gold for her first house and plans to bring young girls from China to help with the milking. I suggested we go into business together but she wants it all.

FIRE! FIRE! The shouts and screams tore through the middle of the night. Fire on the north end of town. The Committee was shouting for people to come quick with heavy boots, blankets and buckets.

Two of the Indian brush huts were on fire and if the wind changes the rest will go up as well and spread to the town.   Sparks were flying into the air and the Indians were trying their best to save the rest of their huts.

Men stomped the flying sparks when they fell on the ground and threw a wet blanket over them when they fell on a hut. A bucket line was formed from the creek and bucket after bucket was thrown onto the burning huts.

After a couple of hours of hard work the fire was out. I went over and signed White Fox. Two children burned but not bad. Can rebuild huts in a couple of days. Everyone has a place to sleep tonight in other huts.

Betsy invited everybody to the Fun House for free drinks. Most were too tired and went home to bed. A few came, downed a drink or two and collapsed on a chair or the hard wooden floor.

During 1848, a Methodist minister came to town now and then to preach. There was no permanent minister or church until a man of God, full of the holy fire, came riding into town in late January, 1849.

He led a dozen followers and they proceeded to buy a lot and with help from the Ladies Guild build the first church of God in Hangtown. Most of the womenfolk were glad, the menfolk not so much. Praise the Lord, they didn't bring a church bell with them.

Harriet went off on an almighty rampage, screaming and waving her arms in the air, tearing into the bible thumping hypocritical fools coming to town, just like the ones who threw her out of St. Louis. Betsy quieted her down. We don't want trouble until trouble comes to us.

Sure, sooner or later the town will get lawyers and judges, politicians and taxes, and the free life will be gone. Best not hurry that along.

More and more stealing going on. Early gold seekers are arriving from around the Horn and across the Isthmus.

Mid-February the Committee met and decided to start whipping anybody caught stealing or claim jumping. The warning was sent around the town, but didn't help much, so the whippings began.

I went once. After being found guilty the man was given the Choice between hanging or whipping. He was tied to a tree and his back stripped. A man chosen for strength and dedication applied the whip.

A good many of the spectators laughed, cheered and made crude jokes as each lash fell. Twenty lashes left the man bloody, scarred for life, and passed out. I never went again.

Ya. Ya. I know we have no jail so it's either the whip, the noose, or let them thieve again.

Business was booming at the Fun House but fights were more common and lawlessness was increasing so we added another bouncer, a good friend of Mike and Johnny. A few more girls drifted in from Sacramento and San Francisco, looking for a little excitement and their pot of gold.

The PLACER EMPORIUM is getting low on merchandise and THE FUN HOUSE is running out of supplies. Time for another trip to San Francisco.

Mike and Johnny, our two bouncers, and Betsy, Molly, and I started out on a sunny day in the middle of March. Charley, the bouncer left behind, will bring 25 mules to the Sacramento dock in 5 days.

We stopped for early lunch next to a pretty little waterfall and heard rustling in the bushes. Three men walked out shouting:

HANDS UP! HANDS UP!

Johnny moved his hand toward his gun but saw the desperados guns swing toward him and thought better of it.

"Ok. We want all that gold you're carrying. Hand over the gold and nobody gets hurt. You give us trouble and sure as hell you be dead."

We girls went poor weak female, shrieked and whimpered. They kept their guns pointing at Mike and Johnny and went for the saddle bags lying against a rock nearby. Bad mistake.

Betsy and I shot two of them and they went down. Before we could shoot the third he turned and shot Molly. Betsy and I fired a second later and his head exploded.

Dancing Deer, Sister Mary Ellen and now Molly. I didn't know how much I loved that little girl. I cried my eyes out over her body. Still warm. So dead.

Life moves on. We had no shovel but found a rock pile just up the creek. Hauled out rocks to make a deep hole, carefully placed her body at the bottom, and piled on the rocks.

                    Goodbye Molly, old friend
                    Fare thee well.

We tossed the three robbers into some bushes for the buzzards and coyotes, and looked through their saddle bags. They are very successful robbers. Around fifty pounds of gold dust. We all got an equal share.

We agreed to tell no one what happened. When people ask, tell them Molly's horse got spooked by a rattlesnake. She was thrown and hit her head on a rock.

We had to move fast to make the steamer whistle. It was not a happy journey down river. I kept seeing piles of rocks. Black and squirming as if they were alive.

Oh my God, what a change. San Francisco is many times bigger. The tents are gone downtown and adobe and brick buildings are built or being build on the Plaza, the center of town.

We went to the Parker House but full up so settled for the City Hotel across the plaza. They have soft mattresses and clean but ordinary sheets. A little disappointed because we heard that the bed sheets at the Parker House are French Silk.

~~~~~~~~~~~~~~~~~~~~

The first picture is the way it looked around 1854 or so. They added all those fancy fences, trees, grass, and paths. The second picture is a little later. The trees have grown nicely and there are flowers.

The BELLA UNION was the best gambling hall in town. It was a wild place, even in the daytime.

~~~~~~~~~~~~~~~~~~~~

I didn't even try to eat dinner and had bad, bad dreams full of exploding heads and piles and piles of heavy, black rocks.

The next morning Betsy and Mike went to buy for the Fun House. The Chinese wine was a big hit, as well as home brewed whisky from a bunch of stills in the hills south of San Francisco. And glasses got broken and bed sheets and linens wear out faster than they should, and we need to double the cleaning supplies.

I bought for the Emporium. Johnny carried the gold. We need as many boots as we can get. They wear out real quick in the mud and water. Most of the other stuff we bought on the last trip has been selling well.

Mr. Munroe gave me a list of the merchants and merchandise from his first trip. I'll visit the gentlemen and bargain away. Oh, and I need to be sure and buy an extra supply of wicks, oil and candles.

Most of all there is a particular gentleman I want to see. Mr. Levi Strauss. He came to town from around the Horn in early 1849 and began to use heavy brown tent canvas to make real tough trousers. Miners in Hangtown told me they wear much better than any others they have ever had.

I found his shop just off the Plaza. A fine looking gentleman with a German accent.

~~~~~~~~~~~~~~~~~~~~~~~~

I found his picture a few years later and stuck it in here. He's older than when I saw him and didn't have a beard. I have heard talk that he came to San Francisco in 1850. I assure you he came in early 1849.

~~~~~~~~~~~~~~~~~~~~~~~~

"Good morning, young lady. How may I be of assistance?"

"I am looking for trousers. I have heard that yours are much stronger and last much longer than most."

"No young lady, they are not stronger and last longer than most. They are stronger and last longer than all other trousers."

"My goodness, Mr. Strauss, humility does not appear to be among your virtues." He laughed.

"Truth sometimes appears as lack of humility, but remains the truth nevertheless. I assume you are considering buying a pair for your father."

"How much do you charge for a pair?"

"Five dollars."

"Oh my, that is quite a bit of money. Would you be able to lower the price if I bought two pair?" My goal is to get him down to 4.00 a pair for all his trousers. $3.75 if I'm lucky.

"For a lovely young lady like yourself, I will be happy to lower the price to $4.75."

"Oh, thank you," and ever so slightly batted my eyes. "So the more I buy the lower the price?"

"Yes. I try to be as generous as possible with my customers."

"I have others who could also use a pair of your wonderful trousers. If I bought 4 would you find it possible to sell them to me for $4.50 each?"

He hesitated, and looked back and forth between me and his trousers.

"For a nice young lady like yourself, of course. I will be pleased to do so."

"Oh thank you so much. You are a very nice man."

"Thank you. May I wrap them up for you?"

"How low will you go if I bought 10 pairs of your wonderful trousers?"

He straightened up and his eyes opened a lot wider and his mouth hung open.

"Are you serious?"

"That depends upon the price you are prepared to give me."

He looked confused and then a little angry.

"You have been playing a little game with me."

"Yes I have. And I am seriously asking how low will you go if I buy 10 pairs of your trousers."

"I believe that $4.25 is a very fair price."

"How many trousers do you have for sale?"

"Why do you ask?"

"May I have an answer to my question?"

"I have about 150 pair at the present time."

"I am prepared to buy them all at $3.50 a pair."

His eyes got bigger. And he got angry.

"I do not enjoy the game you are playing with me."

"It is no game, Mr. Strauss, but I see that for some reason you do not wish to do business with me, so I will waste no more of your time. Perhaps Mrs. Hertz on the Plaza will be a bit more friendly." And started for the door.

"Wait! Let us assume you are not playing a game with me. If you buy all that I have, about 150, I will sell them to you for $4.00 each."

"I am not playing a game with you, Mr. Strauss, and will pay in gold dust. Let us split the difference. I am prepared to pay $3.75 a pair for all your trousers."

"Let me see the dust."

I went outside and got two pouches from Johnny.

"I believe you will find when you weigh out the gold that there is more than sufficient dust to pay for the trousers."

He picked them up and hefted them a few times."

"You have a deal."

Blue Bell is doing very well. As she predicted more and more people are flowing into town, from across the Isthmus, around the Horn, and from other parts of California and Oregon. And you can almost hear the howls of the hoard coming across the plains.

Miners with sacks of gold are proving to be very generous customers.

She bought a lot near the top of a hill north of the Plaza and with just a little more money will begin to build a fine, high-end house.

I told her about our Fun House in Hangtown and we agreed to exchange girls now and then. Variety keeps the customers happy, and there are always girls who lust after a little adventure and gold, or want to return to the refined life of San Francisco.

I asked around but no one heard of Sophia or Bethany.

The *Alta California* had a report on the second page that a Chinese woman, A Toy, has been to court a few times, representing herself, and won settlements from men who tried to cheat her. That smart, tough, graceful China girl is a very fast learner.

Betsy Ann bought a large lot near the top of Telegraph Hill, overlooking the Golden Gate. I bought a few lots not too far from the Plaza, just for fun, and because I got a very good price from a gentleman who suddenly decided he had enough of the rough life and was riding a clipper back to New York the next day.

On the fifth day we rode the paddlewheel back to Sacramento City, loaded the mules, and back to Hangtown.

10
YOU ARE SPEAKING TO THE PROPRIETOR

You pompous ass.

I hate it. I'm beginning to like Johnny a little too much. I don't want some big ape in my life, ordering me around and making me his little woman. Except Johnny's not like that. He's big enough but he doesn't try to control me.

The problem is I just don't want to settle down and raise a bunch of kids, and have my lady friends from the Guild over every Saturday afternoon for tea and tarts, and perhaps a of slow game is whist. I'm free and I'm going to stay free!

Except, just yesterday, I went way out of my way to watch some mothers play with their children.

Early one morning around 2 weeks after I returned from San Francisco, Aokii'aki took my hand and led to a quiet place under a large pine. I bent down and we touched foreheads.

"Khoo'ii and I return to the mountains of my ancestors. I wish to sleep at Two Medicine Lodge. Sopo'aki, come with us."

I looked in her eyes a long time, struggling with surges of conflicting feelings and thoughts. I so want to return with them, back to the vastness.

And, I want to stay and continue to build a life that brings joy and a different kind of vastness. Yes, I like being a successful proprietress of a mercantile exchange, part owner of a house of joy, and a land speculator in San Francisco.

No! Johnny has nothing to do with it. **Nothing!**

"I see your heart has two paths."

"Aokii'aki you are my mother. My true mother. My heart will always be with you and yours will always be with me."

"You have chosen your path."

"Yes."

"Let it be so."

She did not shed tears but I felt her deep sadness and my own. We sat for a long time there beneath the giant pine, feeling love and sadness. I shed tears for both of us. We will not meet again in this life.

She and Hughie left a week later. Osgar went with them.

        Goodbye Aokii'aki, my mother.
        Fare thee well

~~~~~~~~~~~~~~~~~~~~~~~~~

~~~~~~~~~~~~~~~~~~~~~~~~~

I hired Johnny to man the store when I'm not around. Mr. Munroe will return when Aokii'aki sleeps with her ancestors at the place of Two Medicine Lodge.

To keep up the ruse, and continue to protect Sammy and friends, Mr. Monroe has appointed Mr. Beckwourth to take over their management. The notice of the appointment was placed on the town message board outside Mr. Wheatley's establishment.

A couple of weeks after they left I was tending store when a well dressed young gentleman came through the door. He gave me such a funny feeling, as if I had known him a long time.

"Good morning, young miss. Please inform Mr. Munroe that Collis Huntington from Sacramento would like to speak with him."

"I am sorry, sir, but he is not here. He left recently for Blackfoot country."

"Ah. Then may I please speak with the present proprietor?"

"May I ask your business, Sir?"

"No you may not. I wish to speak with the proprietor."

"Sir, your tone of voice is most uncivil."

"My goodness, young miss, you have a very smart mouth."

"Not nearly as smart as your own, Sir. State your business or leave."

"How a rude whippersnapper like you ever got hired is beyond me. I demand to speak with the proprietor immediately!

"You are speaking to the proprietor, you pompous ass."

###********************

That is all I have been able to save. What a loss if I cannot find a way to recover the rest. Her story is only half told.

###********************

############################

I have looked long and have looked hard for a way to recover the rest of Mary's story. No luck.

Now you must become Mary. Allow the next chapter of her life to flow into your night dreams, and into you day dreams. Become Mary as you lie in bed moving into that twilight zone between awake and asleep. Remember your dream journeys as you wake up. Allow flashes of awareness during the day.

############################

If you liked _MARY: a tale of the Wild West of long ago_, I would be grateful if you would write a review. One or two sentences would be fine, or make it as long as you like.

Click here or Google the title and scroll down to the review section.
Thanks.

———————————

BOOKS AND PLAYS
All available from Amazon as eBooks and paperbacks.
Ctrl-Click the blue book names or cover to learn more.

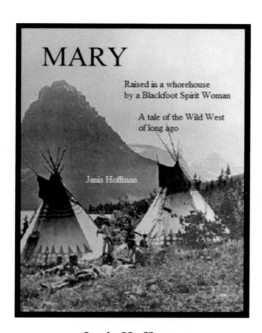

Janis Hoffman

I hold a modestly high level position in government and wish to remain anonymous. I am grateful to Dr. Kinnie for supporting and publishing my books about the unusual lives of Mary and Alice.

Mary

Mary was raised in a brothel. Didn't want that kind of life and jumped on a wagon train when she was 15. Ride with her into the Wild West of long ago. She promises not to lie too bad.

I found her handwritten story in an old trunk I bought at an auction in Walton, NY.

There are many corrections and many notes and pictures stuck between the pages, and the ink and pencil are faded and often difficult to read. I had to guess a few times and hope the language of my guesses doesn't sound too modern, nor done too much harm to Mary's intent.

The name Mary Faraday Huntington does not appear in any of the old records. Whoever wrote the words shamelessly talks about things rarely mentioned in stories of the Wild West.

She describes the way it was long ago in the gold fields of the Sierras and among the soiled doves of San Francisco, not the sugar coated fairy tales of book and Hollywood.

Thoroughly enjoyed this book! I love history and to read it in the first person and in the words of a woman who had "grit" and enjoyed life and adventure was very entertaining.
Kindle Customer

I liked the humor…Indians and whorehouses and the characters were developed to make the story fun. Also loved the old pictures. A fun story and fast read.
Joanhughs

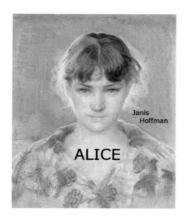

Alice

I love playing poor little orphan girl. Guys are so dumb. I milk them for what I can get. Not for sweet, little snowflakes. Would melt your prissy brain. aCome along if you still have a little wild left. Explore the World of Shadow with me. Lots of fun.

When I was four my mother drove away, leaving me on the sidewalk sobbing and screaming. My father died just before I was born. Not very nice of him but he left me a pile of money and a cute little cottage on Big Sur. Better than nothing.

After blowing 8 foster homes I easily conned my way into UC Berkeley. No problem running circles around the intelligentsia of that fine institution.

———————

I enjoy Hoffman's characters. Their candor and no nonsense attitude.
Jim Parker

My name is Nutmeg she screamed

Schizophrenic mother. Many foster home fails. Children's Garden may be Nutmeg's last chance.

She stood defiant. "Yeah, I fried the goldfish. What you gonna do 'bout it?"

"Not much at the moment," I replied. "What's your name?"

"I am Nutmeg."

"Hmmm, it says here your name is Katy."

"My name is NUTMEG!" She screamed. I wasn't at all sure the sounds coming from this cute little girl in front of me were human, but they were intelligible and very, very **loud.**

Yes, there really and truly was a wonderful place called Children's Garden, in Marin County, California. A special place where broken children were mended. This is the story of Nutmeg. Taken from her schizophrenic mother when very young. She has failed many foster homes.

Written very well. I like that it was told by both the social worker and the child.
Rene Peterson

Absolutely loved this story. Great to see some good in this world come from the foster care system. Good can come from trauma.
Kindle Customer

Great read for those wanting to be foster parents or adopt. This book doesn't sugar coat what life could be like.
Cheryl Carmickle

Ernest Kinnie, PhD
Clinical Psychologist

Over the years I've found many useful psychological concepts and strategies that increase love, joy, creativity, and adventure. My books and website give you the most powerful.

You become one of the creative and adventurous of your generation. Able to bring fresh emotions, thoughts, and actions into the world of humankind.

And have a whole lot more fun.

I've lived a long, full life and had a hell of a good time. Been a professor, psychotherapist, National Park Ranger, soldier, garbage collector, and antique dealer. My proudest accomplishment---at least the most fun---was writing the first fishing guide to Glacier National Park.

Visit my website.
A new Mind Adventure every Sunday morning.
https://www.psychology-zen-freedom.com/

Tigers for your Mind

This book will change your life!
Sure that's click bait. Also **True.**

It took many years working with hundreds and hundreds of students and patients to create this book. The Tigers give you the most useful and powerful psychological concepts and strategies we found. They really will change your life. Come along.

If you are content being a tiger in a cage, no problem. Enjoy. If you dream of breaking out and running **Wild** & **Free**, this is your chance. The Tigers miss their old friend.

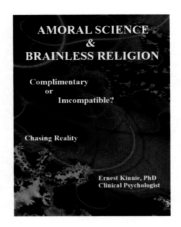

Amoral Science & Brainless Religion

Science and religion are the two major ways western civilization processes Reality. Can both provide valid knowledge? Like to give the definition of "valid" a go? Is there one correct definition or are there many? A Christian would likely consider her experiences in church at least as valid as objective data.

An honest dialogue is often made toxic by a few atheists and fundamentalist Christians who cherry pick history, straw man the other side, and indulge in sad, mean-spirited ad-hominem attacks. The ideologues on both sides trivialize and demonize, and easily knock down the abominable straw men they create---what fun stomping the stuffing into the mud.

In the spirit of John Stuart Mill's *On Liberty*, here's an honest presentation and my take on the strengths and weaknesses of science and religion.

Science

Science is the most valid source of information about humans and planet earth. It has provided wonderful gifts such as longer, healthier lives, and machines that keep us creative and amused.

But best be aware of the influences on scientists that effect the validity of their work---especially in areas that have political, economic and ideological implications. Too much scientific research is junk. Cannot be replicated.

Scientists accommodate the needs and agendas of governments, businesses, interest groups, and themselves---wittingly and unwittingly.

Eric Berne

Yes, you're all grown up now and have to act like an adult. You are reading this book so you have done a pretty damn good job of it. You might have done too good a job and lost the vitality, adventurousness and joy of being a kid.

Every child is an artist.
the problem is how to remain an artist once we grow up.
Pablo Picasso

Don't let growing up shove your kid away.

Margaret Mead found a way around that horror.
I was wise enough to never grow up
while fooling most people into believing I had.

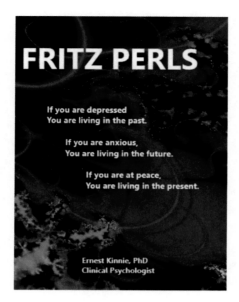

Fritz Perls

Words! Words! Words! They shut one off from the universe. Three quarters of the time one's never in contact with things---only with the beastly words that stand for them.
Aldous Huxley

Yes, you live in the middle of clouds of words. Nothing wrong with that---you would not survive in the "real" world if you didn't mostly use that mode of interaction with Reality.

But now and then blow them away! Let Fritz and the Tigers show you a less filtered world. Your dance with Reality becomes more accurate. The quality of your interactions with others, yourself and the world so much better.

If you're a snowflake or Mrs. Grundy it would be best to pass on by the second and third chapters. Come along if you haven't had your free spirit squeezed or beaten out of you.

Skinner and Pavlov

BRAINWASHING THE MASSES
Politicians and advertisers throw tons of crap at you every day.
They want to make sure you think, feel and act CORRECTLY.
VOTE! VOTE! VOTE!
BUY! BUY! BUY!

Use Skinner and Pavlov to throw all that crap right back in their faces.

You and your friends and enemies also use Skinner and Pavlov every day to influence and control each other. Just part of normal social interaction. You'll become more aware how you do it

In the second chapter use Skinner and Pavlov to better understand your own behavior. You'll be given the tools to control and **further your life**.

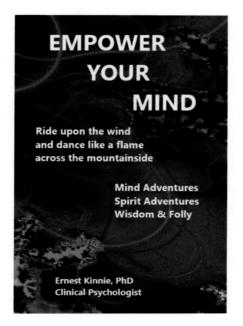

Empower your Mind

Add more joy, love, adventure and creativity to your life. It took many years to find the very best ideas and tools to give you. Yes, and I learned a good many of them the hard way.

You don't just read about the concepts, ideas and tools, you experience them. That's the only way to learn to use them. There are many essential Mind Adventures for you to explore. Then ride the waves of your brain into the wild world of imagination. Take strange trips through space and time. Lots of fun and great fantasy adventures.

You become one of the creative and adventurous of your generation. Able to bring fresh emotions, thoughts, and actions into the world of humankind.

<u>The Magic Theater</u>

Come join the Wizard's Dance.

Every psychological, scientific, political, economic, and philosophical theory has strengths and weaknesses. That realization frees you from the tight box of black and white thinking.

Then there will be a second magic moment. Your **Self** is your primary Reality---your stable point among the shifting realities of the world.

You become one of the creators of your generation, able to bring new thoughts, emotions and actions into your life and into the world of humankind.

And have a whole lot more fun.

Dramatic Moments

FIVE SHORT PLAYS
Five life changing moments

Their last chance
Sammy finds a way
The kiss
Smackin' the monkey
Deep and serious

Dramatic Moments 2

FIVE SHORT PLAYS
Not for snowflakes or finger waggers.

I'm too old for this crap
You want sex?
A cougar's kitchen
Androidian love
High flyers